Handbook of Flexible Circuits

Edited by
Ken Gilleo

Poly-Flex Circuits
Cranston, Rhode Island

VNR VAN NOSTRAND REINHOLD
_____ New York

To the oft misunderstood flexible circuit and to all the technologists who brought vitality to this exceptionally versatile breed of electronic interconnect.

Library of Congress Catalog Card Number 91-27746
ISBN 0-442-00168-1

Published by Van Nostrand Reinhold
115 Fifth Avenue
New York, New York 10003

Chapman and Hall
2-6 Boundary Row
London, SE1 8HN

Thomas Nelson Australia
102 Dodds Street
South Melbourne 3205
Victoria, Australia

Nelson Canada
1120 Birchmount Road
Scarborough, Ontario M1K 5G4, Canada

16 15 14 13 12 11 10 9 8 7 6 5 4 3 2 1

Library of Congress Cataloging-in-Publication Data

Handbook of flexible circuits / Ken Gilleo, editor.
    p.   cm.
   Includes index.
   ISBN 0-442-00168-1
   1. Flexible printed circuits. I. Gilleo, Ken.
TK7868.P7H36   1991
621.381'5—dc20   91-27746
                        CIP

Contents

Foreword

According to the IPC's Technology/Marketing Research Council, in the last 20 years the U.S. market for flexible circuits has gone from 15 to over 350 million dollars. Today, flexible circuity is seen in computer keyboards; in hard drives; in telephones, tanks, and aircraft. Flex circuitry is used with Polymer Thick Film and Surface Mounting, and performs in both benign office environments and harsh automotive environments.

The Institute for Interconnecting and Packaging Electronic Circuits (IPC) formed its first committee to address flexible circuitry in 1962 and issued its first Standard on flexible circuitry in 1965. Since then, IPC has issued Standards on materials, design, and assembly of both flexible and rigid flex applications. Test programs are sponsored to continue gathering information for the industry.

Still more information is needed. Handbooks, such as this, gather information and document industry expertise to be passed on to others faced with the challenges ahead. While the market may not grow as it has from 1970 to 1991, new opportunities and applications for flex circuits will most assuredly be found. We at IPC wish the reader of this book success in creating future markets and meeting future challenges for the use of flexible circuitry.

David W. Bergman
Director of Technical Programs

Institute for Interconnecting &
  Packaging Electronic Circuits

Preface

Flexible circuit technology, whose concept dates back to 1904, was ahead of its time but, unfortunately, poorly utilized. Flex, the original 3-dimensional circuit, has experienced dramatic growth and development over the past decade as its unique features solved a myriad of problems created by the information revolution and the new wave of electronics. This book systematically presents the essential information to allow the reader to fully exploit the extensive array of features and benefits of flex.

The book describes flexible circuit technology in its broadest scope, yet with intricate detail. Subsets, such as Tape Automated Bonding (TAB) and Membrane Switch are included. Materials are discussed in considerable detail, so the reader can develop a good understanding of the limits and useful ranges. Later chapters enumerate various construction options and set forth design guidelines.

Emphasis is placed on methodically guiding the reader to the optimum use of flexible circuit technology and avoiding under-utilization of features or over-stressing the products. Design for manufacturability is a constant theme. Chapter 4 on Design and sections dealing with applications, attempts to orient the reader to 3-dimensional thinking. The prime goal is to transform one from the flat "rigid" world of 2-dimensions to the exciting 3-D "volumetric" space of flex. All drawings in the book were done by the editor, with the help of MacDraw.®

What is a Flexible Circuit?

Flex is an object of beauty to both the eye and to the mind. Its elegance in design proclaims the keen talents of the creators of this extraordinary article and its stunning precision reveals the integrity and dedication of the crafters. Simplicity of concept blended with an extravagance of features is the hallmark of flex.

Thin and transparent, with gleaming metal conduits carrying power and messages, flex links together the world of high technology. Supple and yielding, curving and gliding from point to juncture like a carefree bird, flex interconnects in a quiet and boundless way. As electronic devices become more dense and products become perpetually more compact, the arteries of electrons and waves must also become miniature. Flex will provide unique answers as we escape conventional two-dimensional boundaries of thinking and designing.

Flex, with subtle strength and the ability to take on a thousand shapes by blending and conforming, brings solutions for today and tomorrow. Flex, with unmatched suppleness and compliancy adds the dimension that others forgot: flex adds the third dimension to conquer time and space. This beautiful yet awesomely powerful creation, serves the world in a thousand ways. In slim calculators, powerful computer peripherals, and the modern automobile, flex is there. These circuits gently cradle the modern silicon masterpieces to bring the micro world of electronics to the human interface. We will explore the evolution of this unique and wonderful entity called *flex.*

1

Introduction

Ken Gilleo

Poly-Flex Circuits
Cranston, RI

1.1 DESCRIPTION OF FLEXIBILE CIRCUITRY

Flexible circuitry is a patterned array of conductors supported by a flexible dielectric film. This simple definition for such a specialized, but highly versatile electronic product, may lead one to the mistaken conclusion that flex is merely a cable that bends easily. Flex was originally used only as an interconnect—a convenient replacement for point-to-point wiring. Hence, the old name Flexible Printed Wire (FPW) was adopted. Flex was then, and is today, the premier dynamic cable with an extraordinary ability to withstand continuous flexing for hundreds of millions of flexing cycles.

Today's popular term, Flexible Printed Circuitry (FPC), more closely reflects contemporary usage. Modern flexible circuitry is a bona fide electronic assembly circuit, a superior interconnect, and much more. Flex also has the unique capability of easily incorporating many unique features. Advanced flex can provide such integrated elements as connector pins formed from conductors and TAB (Tape Automated Bonding) chip bonding sites for directly connecting IC's to the circuit. This book will thoroughly explore the myriad of features, functions, and benefits that make flex today's most versatile electronic interconnect medium. Design rules and guidelines will be clearly described. Case histories, with intriguing solutions, will be presented to bring the fascinating story of flexible circuitry together and into perspective. ·

Increasingly, flex is referred to as a Flexible Printed Circuit (FPC) as its use as an assembly medium increases. Surface Mount Assembly (SMT) is an ideal package technology for flex and its success in the 1980's has substantially boosted the use of flex for complex electronic assemblies. The two technologies, flex and SMT, share a mutual synergism with each improving the other's performance. These technologies also share common features, such as size and weight reduction, which are enhanced in the combination. Much of Chapter 9 on Assembly will be devoted to SMT-Flex.

One of the most important attributes of FPC is the thinness of the dielectric, which ranges down to 12.5 microns (1/2 mil). This thinness provides unique properties for solving some of the problems brought on by the advances in IC's. Some of the latest applications for flex do not even utilize the flexibility of this product, but rather the properties of the thin, high performance dielectric. In fact, flex is often rigidized for use. Application to metal backers provides excellent heat management, for example. The compliancy of the thin dielectric also provides strain relief for components during thermal cycling. Solder joint fatigue, associated with leadless packages, is almost never a problem with flex as a consequence of the dielectric's compliancy.

We will explore these properties that make flex the most versatile and unusual form of electronic circuit and interconnect in use today. Flex, the original 3-D circuit is becoming the designers delight for answers to density demands, circuit integration, and 3-D concepts. This book will provide you with all the data, criteria, and basic ideas needed to successfully design, specify, and use flexible circuitry.

1.2 HISTORY OF FLEX

The very earliest circuits were made on flexible substrate. In fact, flex is the original printed circuit! This is not really surprising because the early printed circuit was simply a wire replacement. The concept of wire replacement with a flexible conductor array goes back to nearly the turn of the century. In 1904, Frank Sprague asked Thomas Edison if he could come up with an alternate to electrical wires. Sprague wrote, "Dear Edison: As the boss chemist—tell me something. I want to draw a line on a piece of linen paper which will be a fairly good conductor." Edison replied by letter, "Ink thickened with gum and dusted over with electrotypers grafite—ditto gold lead. Strong solution nitrate silver and reduced to silver. . . ."[1] There is, unfortunately, no evidence that anything ever came of these ideas. These concepts are viable today.

There are many early patent concepts, dealing with flexible laminates, that could have been used to make circuits, although they were proposed for other uses. In 1916, Novotny applied for a patent describing the production of a composite consisting of metal foil, adhesively bonded (phenolic resin) to flexible fiber board.[2] The resulting laminate, copper foil bonded to pliable fiber board with Bakelite resin, although intended for use as a printing plate, could readily have been made into a flexible circuit by selective etching.

Copper etching was already well known in the printing industry at the early beginning of what is still the electronic revolution. By the early 1920's, fine line resolution, half-tone printing plate methods were in use. Bassist described a printing plate making process where copper was coated with photosensitive resist. The dried emulsion was then light-exposed through a photo mask. The resist was finally developed by washing away unreacted emulsion. The remaining emulsion served as an etchant resist.[3] In a short time, the emerging electronics industry would apply these concepts to circuitry.

In 1925, a patent was issued for a method of making electrical circuits.[4] The process consisted of placing a stencil mask over soft metal sheet. Copper was electroplated onto the exposed metal. The mask was removed and the copper was stripped from the soft metal by melting. The copper component (coil) or circuit was then bonded to a dielectric sheet. One alternative described within the patent included printing conductive paths onto a supporting dielectric followed by electroplating. ". . . printing a conductive layer of material upon a supporting surface in the form of an electrical circuit, and electroplating . . ." This appears to be the first patent describing a practical printed circuit.

Two years later, Seymour received a patent which described the manufacturing of a flexible printed circuit.[5] Seymour claimed ". . . the use of a flexible, relatively thin, pliable body in the nature of an insulation and capable of being made to assume various forms or with the different parts thereof in differently related positions." The inventor clearly understood three dimensional electrical interconnect, 60 years before 3-D circuits were invented again. The flexible dielectric was gutta percha (a natural rubber; coagulated exudate from Malayan sapotaceous trees used as an electrical insulator),and the conductors were electro-deposited copper or silver. Later patents, especially those dealing with coils and transformers, also described flexible circuitry.[6] Flex was the obvious way to go, then and now.

By the time Dr. Paul Eisler, often referred to as the father of printed circuitry, came onto the circuit scene, there was substantial prior art. Eisler was issued a number of circuitry patents in the 1940's. However, the patents did not withstand the legal test in a fierce court battle. Key Eisler patents were declared invalid in 1963. A crucial patent sited against Eisler was the Seymour flexible circuit invention.[7] By the time Eisler got his famous printed circuit patents, the flexible circuit had already been conceived 40 years earlier and patented 17 years prior. Perhaps, not a legitimate father of printed circuitry, Eisler made substantial contributions to our industry.

1.3 ELEMENTS OF A FLEXIBLE CIRCUIT

1.3.1 Dielectric Base Film

Let us look more closely at the construction of flex. The dielectric film is the most important component of the flexible circuit. Today's materials have many of the best electrical, mechanical, and chemical properties that can be engineered with modern polymer chemistry. The dielectric film is very thin, ranging from only 12.5 microns (0.5 mils) to about 125 microns (5 mils) for polyimides. Low cost polyester film spans even a greater range with thicknesses of 10 mils or more. A more common thickness range for pennies-per-square-foot polyester is 75–125 microns (3–5 mils). Although numerous electronically and mechanically suitable flexible films are available today, two materials have gained prominence: polyester and polyimide continue to strongly dominate the flex industry. However, special composites, which provide bendability, have continued to make modest inroads, especially where 3-D designs are employed.

Very thin and moderately flexible epoxy-glass materials became available in the late 1980's in the U.S., Europe, and Japan. These materials are slowly being adopted by the flexible circuitry industry. Materials under 8 mils in thickness can even be processed by contiguous web techniques. FR-4 product, of only 100 microns (4 mils) thickness, is being used to produce a form of TAB/flex circuitry. These new forms of flex are expected to become increasingly popular for non-dynamic flex applications. The glass-resin combination cannot withstand continuous flexing. However, the utilization of dielectric, once used only by the rigid board industry, means that the previously clear distinction between flex and rigid has blurred considerably.

Polyester film, typified by Mylar®[1] is the low cost, high volume leader. Over two dozen companies now supply polyester film, however. At pennies per square foot, polyester films have provided the world's lowest cost circuits. Polyester, an ideal dielectric material from every criteria except high temperature stability, is widely used where no soldering, or very limited soldering is required. New localized heat soldering techniques and low temperature cure polymer solders, will undoubtedly boost the use of polyester-based flexible circuitry even higher. Large film producers in Japan and Europe continue to improve the intrinsic performance and the film formation properties of polyester. The combination of improved materials and more compatible component bonding methods will ensure the continued wide spread use of polyester. The automotive industry alone consumes over one million pounds of polyester film for flexible circuit applications.

Tens of millions of automotive instrument cluster (dashboard), calculator, printer, camera, keyboard, and miscellaneous consumer electronic circuits are made with polyester base. These products are used for dynamic, or continuous flexing, flex-to-install, and some non-flexing applications where cost savings is important. The fact that polyester film softens below soldering temperature restricts, but does not prohibit its use in solder assembly. Newer soldering techniques like laser and precision hot bar, make moderately complex soldered as-

---

[1]Mylar is a registered trademark of Du Pont

semblies practical, however. These soldering techniques, as well as newer solderless methods, for polyester are thoroughly discussed in Chapter 9 on Assembly.

Polyimide is the premier high performance dielectric of flexible circuitry. Its familiar transparent orange-brown color is found in a large variety of products requiring higher temperature use or extensive soldering. Virtually all military circuits require polyimide in their flex designs. The extreme temperature performance of polyimide (decomposition temperature of over 800°C) makes soldering relatively simple. Properties of the original polyimide, Kapton® and the newer copolymers, are compared in Chapter 2 on Materials.

1.3.2 Conductors

Flex is unusual in that it employs a wide spectrum of conductor materials. A variety of copper foil types and thicknesses are commonly used. These can be divided into two basic categories, electrodeposited (ED), and rolled annealed (RA). Nearly a dozen grades of copper foil, with specific performance requirements, are recognized by the flex industry. Other metals, including resistance alloys, are used for special applications, such as heaters. Additionally, Polymer Thick Film (PTF) conductors are finding increasing use. Conductive inks made by filling polymer binder with silver, copper, or carbon particles, are formed into conductors by simply printing onto the dielectric and drying the ink. This fully additive technology uses a very simple process, but is restricted to relatively large trace widths and spaces.

Rolled annealed copper is almost always used where long-term continuous flexing will occur. RA copper is capable of being flexed for hundreds of millions of times. Many non-dynamic applications also specify RA, but the advent of high ductility ED copper is reducing the RA monopoly on flexing applications. RA copper is more expensive and requires adhesion promoting steps for combining with dielectric to produce circuitry laminate.

ED copper is widely used on flex-to-install applications and where limited flexing will occur. ED is preferred from a cost (lower), availability, and manufacturing (easier bonding) consideration. Newer ED coppers, with small grain structure and the ability to anneal during heat pressing of the circuit, is beginning to find use in less critical dynamic flex uses.

Polymer Thick Film (PTF) conductors are extensively used to make low cost membrane switches and, more recently, for simple circuitry. PTF conductive ink consists of conductive particles, such as silver flake or carbon, dispersed in a polymer resin binder. Future materials may employ copper particles and perhaps, intrinsically conductive polymers. The PTF ink is routinely applied to the flexible substrate by screen printing. The entire process consists of simply printing and hardening the ink. Flexible base film, usually polyester, is widely used with PTF to manufacture very low cost circuits, primarily for consumer electronics use. In addition to low cost, thin, flexible polyester permits the process to be run in a continuous roll for extreme efficiency. Single-sided, double-sided, and multilayer circuits are now made with PTF inks. Chapter 10 on Polymer Thick Film Flex gives a short history and a preview of what is to come in the emerging flex PTF technology.

1.3.3 Adhesives

Adhesive-based laminates are made by adhering copper to flexible dielectric using special flexible adhesive. Although adhesiveless clads are gaining in popularity, adhesive laminates are still the work horse materials. The flexibility requirement puts a severe demand on the adhesive. The many excellent adhesives used for rigid board cannot be used as is. Epoxy, for example, is much too brittle without substantial modification. Acrylic adhesives, once the standard of the industry, are slowly being displaced by lower thermal expansion, higher stability materials. High temperature polyimide adhesives are gaining in popularity due, in part, to their low thermal expansion. Many adhesive systems are mixtures of two or more types of polymers. Polyester-epoxy compositions are popular for high performance. The epoxy provides thermal stability and chemical resistance while the polyester introduces flexibility. Adhesive classification, performance and applications are thoroughly covered in Chapter 2.

1.3.4 Platings and Finishes

Conductors may be treated with antioxidants, also called anti-tarnishing agents, to keep the copper surface relatively bright and solderable. Today, many flexible circuits are used for surface mount assembly because of the high reliability afforded by flex. A highly solderable surface is virtually a requisite for SMT. Nitrogen base organic compounds can react with the surface of copper to form compounds that are stable at ambient temperatures, but decompose during soldering to provide a clean metallic surface. Benzotriazole and imidazole are popular treatments for bare copper circuits. Thin coatings of rosin can also be applied where a longer shelf life is required.

Some requirements for performance are best provided by plated metal finishes over the copper conductors. Solder is the most common coating and has become an industry standard in many applications. Polyester-based flexible circuits are almost always solder-plated if they are going to be soldered. A solder finish provides a surface that can be rapidly soldered, thus reducing the time that the temperature-sensitive polyester is exposed to heat. Some assembly processes, like hot bar reflow, use only the solder that has been plated onto the circuit. These methods will be covered in Chapter 9 on Assembly.

Solder coating may be applied by any of several processes. The primary methods are tinning—applying molten solder, reflowing solder paste, or electroplating. Electroplating provides the best thickness control and can be used on heat-sensitive films like polyester. Solder not only provides a durable, chemically inert surface, but it offers the optimum in enhanced solderability. Any small amount of oxide that may form on the solder, is easily displaced when the solder is reflowed.

Calculator circuits and other low component count commercial assemblies, are often solder plated to provide the necessary solder for component bonding. Hot bar bonding is popular for a gull wing style of surface mount IC package. The heating bars must be able to make direct contact with the component leads and the gull wing shape is well suited. Components may also be bonded to solder plated copper by focused infrared (IR) heating and laser soldering. The solder coating on the circuit supplies sufficient material to form the necessary polyester circuits because the localized heating is compatible with these more temperature sensitive kinds of circuits.

Solder plate is also used to form durable contact pads. Edge contacts, found in flexible cables, are typically solder plated. Key pads for switch contacts can also be plated with solder although carbon ink is sometimes printed over the solder to form a more predictable contact. Solder finish is being used increasingly to make contacts for mating to flat panel glass displays. Elastomer connectors and anisotropic conductive adhesives are used to mate Liquid Crystal Displays (LCD) and other flat panel devices to solder-plated flex. Carbon PTF ink should be printed over the solder plated connection zone.

A thin gold finish is used in applications where minimum electrical resistance and maximum reliability are required. Gold is an inert metal and nearly immune to chemical reactions, including oxidation. The expense of gold dictates that the thinnest coatings be applied. Several gold processes are available. Immersion, a metal displacement reaction, produces the thinnest and least durable. The gold-forming reaction only continues while there is exposed copper at the surface. Electroless gold and electrolytic gold are more common methods. Gold plating can be adjusted to give different hardnesses for specific applications. Gold has a high affinity for copper and will diffuse into the metal unless a barrier is placed between the two metals. Nickel plating, an ideal barrier, is almost always applied prior to gold plating to prevent gold migration. Conductors that will be used for TAB or wire bonding—processes that are covered in Chapters 8 and 10—must have specific thicknesses of gold and underlay nickel. These bonding processes also require precise hardness ranges for the gold.

Both gold and solder can be applied selectively to copper conductors. Temporary plating masks are often used to prevent plating where it is not desired. Printable masks may be used that can be peeled or stripped after plating. A series of masking and plating sequences can be used if more than one type of selectively-plated metal is required. Selective solder and gold plating on the same circuit are often used to achieve the necessary functions. Soldering to gold-plated conductors is usually avoided because brittle intermetallic compounds form. The selective plating approach permits application of each metal only where needed.

Solder is sometimes used as an integral part of the patterning process. Plating resist can be applied to the copper laminate, followed by solder plating. The resist is then stripped away and the circuit etched. The solder is an excellent etch resist for alkaline etchant. The solder can be left in place as protection or for solderability. However, this common, low cost process, leaves the edges of the copper traces exposed. Where this is undesirable, the solder may be reflowed or leveled with heat. These plating, etching, and leveling processes are covered in Chapter 7 on Manufacturing.

1.3.5 Cover Coats, Cover Layers, and Solder Masks

Many, but not all, flexible circuit designs require a protective insulation layer over some of the conductor array. The cover may be used to protect the conductors from the environment, insulate electrically, or to exclude solder in all but the assembly region. Two basic types of covers are used. One is cover layer, an adhesively bonded film, usually similar to the base circuit film, that is precut and then heat bonded in place. The other is a broad class, called cover coat, or solder mask, depending on the intended use. The most common cover coat is screen printed liquid that is hardened thermally or by UV radiation. The print and harden cover coats are more economical than cover layers, but performance can be lower. A more recent class of cover coat/solder mask products are photoimagable, where "light" defines the pattern of openings. Higher density requirements of surface mount are demanding the use of photoimagable solder masks.

Cover layer film is used where dynamic flexing is involved, very high temperatures are expected, or harsh environments are encountered. A common rule, especially for dynamic flexible applications, is to make the cover layer identical in composition and thickness to the base dielectric. This places the conductors in a neutral axis where bending places equalized stress on the conductor. For example, a circuit made with 2 mil polyimide would be covered with 2 mil polyimide cover layer. The cover layer adhesive is usually similar to the base film adhesive but with characteristics optimized for covering.

The lower cost print-and-harden cover coat method can be used in most nondynamic flex applications. However, the higher permeability of many cured-in-place materials may preclude their use in some applications like automotive under hood.

Screen printing is virtually the only process used for selective application of these dielectric liquids because thickness of up to 50 microns (2 mils) can be accurately applied. Radiation curing with ultraviolet is an increasingly popular process, especially for continuous roll processing.

The advent of SMT has placed greater demands on the tolerances of cover coats and solder masks. Routing of traces between mounting pads requires a finer resolution than screen printing or cover layer die cutting can provide. Photoimagable liquids and dry films offer the requisite resolution capabilities. Photoimagable liquid or dry film can be applied to conductors, exposed to UV through a mask and then developed into openings with resolution of under 50 microns (2 mils). The two basic systems are "expose wet" and pre-dry before contact exposure. The expose wet method requires collimated light because off-contact exposure is required. The pre-dry approach allows the circuit to be processed on traditional contact exposers. Although this technology has been in place for rigid boards for many years, flexible versions have been slow in developing.

1.4 TYPES OF FLEX

1.4.1 Single-Sided Flex

Single-sided flex, consisting of conductors on only one side of the dielectric, is commonly used for commercial and military circuitry. The single-sided product can also offer the lowest cost and simplest processing. Conductors can be made of electrodeposited or rolled and annealed copper for more demanding applications. Adhesiveless products can be made both additively and by subtractive etching. Vacuum coated copper on dielectric film can be masked with resist, and the conductor pattern plated up to the desired thickness. The mask is removed and residual copper is etched off in this semi-additive process. The adhesiveless products will be initially single-sided due to the difficulty of producing double-sided vacuum-coated materials.

All dynamic flex circuitry is single-sided, at least in the flexing zone. This construction allows the copper to flex with minimum stressing and maximum life. The copper must be placed in a neutral axis, between the dielectric base film and a coverlayer of similar composition and thickness to the base film. This means that all continuous flexing circuits, like printer and disk drive types, are single-sided, even if there are high density requirements. The single-sided requirement for dynamic flex, has pushed fine line down to 50–75 microns (2–3 mils) for this type of circuit. Rolled annealed copper must be used for maximum flex life. Such dynamic flex circuits are capable of hundreds of millions of flexing cycles. More recently, high ductility electrodeposited copper began finding use on less demanding dynamic applications, like portable CD players.

The lowest cost conductors are made with Polymer Thick Film (PTF) inks, see Chapter 10. Nearly all of the calculator circuits, in the world today, are made of PTF carbon or silver ink on thin polyester flexible film. There is, however, a trend toward double-sided products, even for simple calculators. These processes now allow PTF to be fabricated as double-sided, with printed through holes, or multiple layers, using successive printings of conductor and dielectric.

1.4.2 Double Access or Back-Bared

There are many applications where a single-sided circuit can handle all of the power and density needs, but electrical connections must be made from both sides. The simple remedy is to place access slots or windows in the dielectric. This is usually accomplished by special manufacturing processes beginning at the lamination stage. The base dielectric is coated with adhesive, and the access areas are punched out. The base film and copper foil are then laminated together. The circuit pattern must be registered with the access windows on the back. This all adds difficulty and expense because the laminate must be made in a custom process and back-bared copper must be protected from etchant.

Post windowing processes are also available. Laser milling of both polyester and polyimide have been demonstrated, but not widely practiced. Excimer lasers, which remove organic material by photo decomposition, promise much for the future of machining of polyimide, however. Some polyimide films can be chemically milled with a strong base solution. This method is being used commercially for both flex circuitry and Tape Automated Bonding (TAB), which is a specialized form of flex.

1.4.3 Double Sided

Density demands inevitably exhaust the routing limits of a single conductor layer circuit. A small density increase can be obtained by added jumpers in the form of metal "staples" or printed PTF conductor ink "cross-overs." When surface mount is being used, "zero" ohm resistors can be used as cross-overs. However, a double-sided circuit design is the next full level of density increase. Conductor layers are placed on both sides of the dielectric base film. Circuit patterns are then formed from each conductive layer. PTF ink can also be used on one or both sides. PTF processing also allows layers of conductors, separated by printed dielectric, to be applied to one side of the base film. This provides a double-layer circuit, but with conductors all on the same side.

Most double-sided circuits are designed so that interconnecting conduits are formed to selectively link the conductors on the opposite sides of the dielectric film. Many processes have been developed for top-to-bottom interconnection. Rivets, soldered pins, press fit pins, and solder filled holes have been tried.

However, the Plated-Through Hole (PTH) process remains the most widely used interconnect method. The PTH process involves fabricating holes through the two conductive layers and a dielectric film as a first step. The holes are next sensitized for plating. Electroless copper, followed by electrolytic copper plating, is used to form the PTH, or copper "barrel" interconnect. Several different masking and plate-up sequences can be used, and these will be covered in Chapter 7 on Manufacturing.

1.4.4 Rigidized Flex

Flexibility is highly useful and an obvious key reason for the popularity and utility of flex. Yet, there are situations where applications are best served with a mechanically stabilized or "rigidized" region of the circuit. Backer boards or stiffeners, as the non-circuit supports are called, serve to rigidize part or all of the flex circuit. The term "selectively rigidized" is used to denote designs where only a certain portion of the circuit is backed. Common rigidizing materials include low cost composites, plastic sheets, and even metal. The backers may have access holes but *not* circuitry. Bonding is usually accomplished with heat activated or pressure sensitive adhesive film.

1.4.5 Multilayer and Rigid Flex

Military flex is highly skewed toward multilayer (ML) constructions, although ordnance circuits, for example, are often single layer. All military multilayer is presently made with polyimide. Special adhesives with the lowest possible thermal expansion are being used increasingly. During the 1980's, a commercial ML market has been developing in the U.S., Japan, and Europe. High density disk drive circuitry now requires more complexity than can be provided with double-sided circuits. Laptop computers are also using multilayer flex.

Multilayer flex is constructed by adhesively bonding diclad circuits together in a laminating press. The laminating process is similar to that used in hardboard multilayer except that a vacuum press or autoclave may be used to minimize voids. Holes are then drilled through all the bonded layers in the pressed stack up. The holes are typically cleaned and prepared by chemical or plasma desmearing and etchback. These processes remove drill-smeared adhesive and also expose copper to allow good connections to be made during subsequent copper plating. Copper is plated inside the holes after sensitization which promotes electroless copper plating. The resulting Plated-Through Hole (PTH) copper "barrel," fortified with electrolytic copper, interconnects the individual conductor layers together. Figure 1.1 shows a typical cross-section. The PTF process is covered in Chapter 7 on Manufacturing.

The combination of adhesives and dielectrics usually has a temperature of

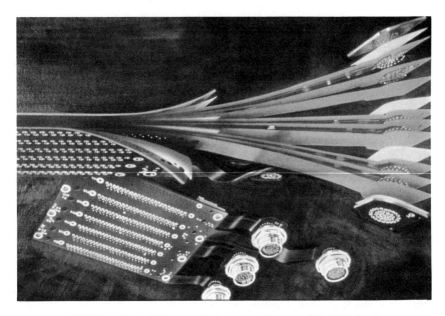

FIGURE 1.1   Multilayer exploded view. (Courtesy of Sheldahl, Inc.)

expansion significantly greater than copper. This difference in Thermal Coefficient of Expansion (TCE) causes stress during thermal cycling. The dielectrics expand to a greater extent than the copper barrel during heating, which can elongate and even destroy the interconnect barrel. The TCE differential limits the total number of layers that can be used for a reliable ML circuit. The difficulty in building a reliable multilayer circuit increases dramatically with layer count using this traditional process with standard materials. Layer count is typically limited to below thirty. This is lower than for hardboards which use inorganic fillers, thus a reduced thermal expansion. Use of expansion-reducing fillers cannot be applied to flex because flexibility and flexural fatigue life is reduced.

One solution is the employment of lower expansion dielectric construction materials. The ideal situation is a perfect match between dielectric and copper. Lower expansion adhesives and dielectric films, which became available in the late 1980's, will eventually allow an increase in layer count. Adhesiveless materials are also being introduced as a solution to thermal expansion problems. This approach, however, requires significant processing changes and may not become fully implemented in the near term.

Composites of flexible and rigid layers can be combined to produce rigid-multilayer circuits, also called rigid-flex. This construction should not be confused with rigidized flex, which has a non-electrical backer. Common designs use rigid in one or more localized areas so that flexible junctions are provided between assembly or other high density zones. Two areas may be rigidized that are joined by individual flexible layers to produce a high density and flexible construction known as a bookbinder circuit. Such designs are popular for military circuits, especially avionics where space is at a premium. However, commercial products are also being designed as rigid-flex. High density disk drive circuits, for example, are using a rigid-flex design, but with lower layer count. Chapter 3 on Constructions covers multilayer and rigid-flex in considerable detail.

A more recent process, called Z-Link™, offers a simpler way of making multilayers. Double-sided circuits with plated-through hole interconnects, are electrically and mechanically mated using anisotropic conductive adhesive. This new process is expected to provide a reliable lower cost circuit with ease of manufacturing and versatility in design for commercial ML. No stress is accumulated because individual layers are "glued" to adjacent layers. This construction concept could make very high layer count possible. Mixed layers of different dielectrics may also be possible for further cost reduction or increased freedom of design.

During the late 1980's, "plated post" multilayer circuitry re-emerged. Early in the development of multilayer flexible circuitry, an interconnect method, consisting of plating-up solid copper posts, was tried, but was quickly outpaced by the PTH approach. The solid post provides greater strength than the PTH connection and is not as prone to stress defects.

1.5 COMPARISON OF FLEX TO RIGID CIRCUIT BOARDS

We need to further recognize that flexible circuitry is much more than a circuit that bends. Flex is the most versatile interconnect medium available: it is the original and the most important 3-D circuit. No other electronic interconnect comes even close in the variety of uses. What makes flex so different from rigid?

Rigid board, unless injection molded into a special shape, is a single plane medium. Hardboard exists essentially in a two-dimensional environment that is bounded by area constraints. The ability to configure flex into multiple planes means that its boundaries are volumetric space. Efficient use of volume instead of just area makes flex essentially the highest density medium available. Hardboard can only take advantage of volume by using inter-plane interconnects. This is only done on a limited scale and not too efficiently. Personal computers use mother/daughter board arrangements to move the rigid boards into two planes. Supercomputers and a few costly military assemblies stack a large number of rigid boards into an efficient volume. Flex, on the other hand, accomplishes this feat everyday in a thousand products.

We can look at flexible circuitry's virtue of occupying multiple planes as a self-integrating interconnect. Two hardboards mated with a single connective cable, can be positioned into two different planes. We could keep adding boards and cables to get $N + 1$ number of planes where $N =$ number of interconnects. A single flex circuit can be positioned into any number of planes only limited by size. Flex serves as the interconnect as well as the circuit board. The key concept is that a continuous conductor array is patterned into the desired circuit. No plugs, cables, solder, or joints degrade the signal or reduce reliability. This makes flex the most reliable means of interconnect for multiple planes. Figure 1.2 shows a multiplane flexible circuit.

Flexible circuitry is also the thinnest product in common use. Thinness is becoming increasingly important for a number of reasons beyond simply creating a low profile circuit. Because flexible circuitry is made from extruded or cast film, it will remain the thinnest circuit within the electronics industry. Dielectrics are being used that are thinner than a human hair. A minimum dielectric thickness is very important for heat management. Organic dielectrics have relatively high thermal resistances, about 1 watt/meter $\cdot$ K, compared to many ceramics. Heat transmission through the dielectric is inversely proportional to

FIGURE 1.2   Multiplane flexible circuit. (Courtesy of Sheldahl, Inc.)

thickness. Flex at up to 1/30th to 1/60th the thickness of hardboard, therefore transmits heat at 30 to 60 times the rate, all other things being equal. Designs that take advantage of thermal transmissions, often specify bonding the flex to metal. This is much less practical or useful with hardboard.

The thinnest of flex also provides a key property for assembly—intrinsic strain relief for component junctions. Although surface mount devices provide a host of valuable benefits, their disadvantage is the amplification of thermo-mechanical strain. Without device leads to take up the movement caused by differential thermal expansion, solder joints are stressed during each thermal excursion. Solder joints can fail after a few hundred thermal cycles on hard board. This is fortunately not the case with flexible circuitry. The thin, compliant dielectric base film easily bends during thermal cycling of flex SMT assemblies. The stress normally associated with thermal cycling is absorbed and dissipated by the compliant flex. The solder joints are protected from fatigue stressing, which commonly occurs with all but special hardboards. This important feature makes flex the most reliable circuit substrate for surface mounted components. SMT, with similar attributes of reduced size and weight, is the ideal assembly technology for flex. The mutual synergism between flex and SMD, will be thoroughly covered in Chapter 9 on Assembly.

Flexibility also provides an extremely important manufacturing process feature and advantage. Flexible circuitry can be produced in continuous web fashion. The process, usually called roll-to-roll, allows a large roll of laminate to be imaged, patterned, and plated, all in roll form. Both single- and double-sided circuits are produced by continuous roll processing. This provides important handling efficiencies and improved process control. The roll-to-roll manufacturing method has been in use in the United States for over 20 years and was pioneered by Bell Labs and Western Electric. Several continuous web flex factories are in production around the world. See Figure 1.3. Most new flex plants, particularly in Japan, are being set up for continuous roll.

Another important ramification of roll processibility is its impact on assembly. Flex circuits can be produced and used in a continuous roll. They have been and are being used in continuous roll form. During the 1960's, calculators were assembled from continuous rolls of flex using an early form of surface mount. Dual In Line (DIP) packaged IC's were configured into gull wing SMD's. These were hot bar soldered onto rolls of calculator circuits, which were finally blanked out. Although the process was abandoned because of early equipment shortcomings, it proved the viability of the concept. Today, the continuous roll assembly process is again being used.

Tape Automated Bonding (TAB) is a specialized form of continuous flex. TAB consists of a fine line circuit on a flexible base film. The dielectric and conductor materials are typically the same ones used for standard flex. Two features distinguish TAB from ordinary flex, however. An opening, or "win-

FIGURE 1.3   Continuous web flexible circuitry plant. (Courtesy of Sheldahl, Inc.)

dow,'' is fabricated in the dielectric so that thin copper traces, called ''fingers,'' can be cantilevered over the opening. These thin, wire-like conductors, can then be thermo-compression bonded to an IC. The second special feature is a series of sprocket holes along each edge of the tape: this is used to advance and register the substrate for bonding. TAB has been used as a continuous roll, or reel, bonding substrate since the 1960's. The product has the basic appearance of a roll of photographic film, with its edge sprocket holes. Although TAB is not always recognized as a subset of flex, the product was originally produced by the flex circuit industry. Many flex manufacturers are beginning to offer TAB or TAB integrated into conventional flex. TAB and other integrated interconnects of flex will be covered in Chapter 8 on Integrated Features.

1.6 OUTLINE FEATURES AND ATTRIBUTES

We have covered a number of the features and benefits of flex. This section will serve to consolidate and review those that are key attributes.

1. Flexibility: the property of being easily and repeatedly flexed permits three important uses:

a. Flex can be bent to the necessary profile to install into a product. This is referred to as flex-to-install application. It involves a one time flexing, or a limited number, if repairability or component replacement will be required. These applications do not require special copper foils. Calculator circuits are a good example.

b. Flex can be conformed to an enclosure or housing. This is really a 3-D circuit application. Standard ED copper is used in general applications. Automotive instrument cluster circuits are a popular example.

c. Flex, when properly designed with the right materials, is capable of being flexed continuously and virtually indefinitely. This application is called dynamic flex. Care must be taken to select a ductile copper and to balance the materials. Design rules must be applied to the pattern geometry. Floppy disk, hard disk drive, and printer circuits are high volume products using the dynamic flex feature.

2. Compliancy: the property of bending under stress conditions in a way that reduces or completely eliminates strain on components. An excellent example of strain relief compliancy is the surface mount flex assembly. Flex provides outstanding resistance to solder joint fatigue by dissipating mechanical forces during temperature cycling. The thin flex bends during differential expansion and contraction so that force on the solder joint is minimal.

3. Thinness: the property of having the processing and functional properties to be manufactured and used in very thin format. Thinness permits unrestricted bending and flexing. Heat can be readily conducted through the dielectric to a heat sink or to air. Insulation displacement connectors can be installed. Other attributes and design features derived from the thinness will be covered throughout this book. It will suffice to say that the unique thinness of flex provides more benefits than any other characteristic.

4. High Temperature Performance: the ability to operate as a circuit or interconnect at elevated temperatures. Polyimide film and adhesive, with its unprecedented high temperature performance, is the basis for outstanding high temperature limits of flexible circuitry. Operating temperatures in excess of 125°C are documented. Performance is being pushed to 150°C and beyond.

Often, there is adhesive degradation before the base film is altered by heat. New adhesiveless copper polyimide clads are expected to boost temperature performance to new levels.

5. Etchability/Machinability: property of allowing the dielectric base film or cover layer to be selectively removed. Many designs for flex require openings in the dielectric film. Double-access, for example, requires removal of base film to form access openings. Although the base film can be mechanically patterned prior to laminating copper, versatility is substantially increased by the ability to remove dielectric material after the conductor pattern is formed. Polyesters and polyimides both respond to laser machining. Many polyimides can be chemically milled. Strong bases depolymerize and dissolve the polyimide backbone. Adhesiveless copper polyimide products are excellent candidates for this process. Adhesiveless TAB, called 2-layer TAB, is fabricated by chemical milling.

6. Interconnectability: the property of providing a convenient electrical pathway between locations in different planes or in relative motion. Flex designs, which provide interconnections, continue to be an important segment of the market. There is a trend to add electronics to interconnect designs. Surface mount and TAB packages permit small, lightweight, yet powerful electronics to reside on interconnect cables. Disk drive assemblies for hard drives are a good example.

7. Size Reduction: the ability to reduce total circuit area or electronics volume over conventional rigid boards. Flex can eliminate cables and connectors to reduce size and increase reliability. The ability to roll, bend, fold, and otherwise change the shape of a flex circuit provides considerable size reduction for the system. SMT assemblies, for example, can be folded over to reduce area by 50%. Accordion-like designs go even further in size and volume reduction. The foldover edge card, used as a plug-in component on VCR's, is one of many examples. See Figure 1.4.

8. Weight Reduction: the ability to reduce circuit and assembly weight compared to conventional circuitry. Flex, by virtue of its thinness, is intrinsically weight-reducing. Elimination of connectors, cables, and plugs also impacts weight reduction. However, the total flex system generally represents significant size, component and, therefore, weight reduction. The credit card is a good example of miniaturization, low profiling, and weight reduction.

9. Simplification: reduction in total system complexity. A single flex circuit may replace numerous point-to-point wires. A more complex flexible circuit configuration can replace literally thousands of wires in different plans for substantial simplification. Because the flex is arrayed in a fixed pattern, wiring error is virtually eliminated. Military high density multilayer interconnects are a good example of simplification.

10. Volume Efficiency: effective use of 3-dimensional volume by the as-

FIGURE 1.4   Foldover flex on aluminum backer. (Courtesy of Sheldahl, Inc.)

sembly. Flex makes the utilization of volumetric space relatively easy. The circuit can be rolled and folded to fit a limited space. Automotive sensors, housed in compact mounting housings, are a recent example of volume efficiency.

References

1. Cadenhead, R. L. and DeCoursey, D. T., The History of Microelectronics—Part One. *The International Journal of Microelectronics*, Vol. 8, No. 3, Sept. 1985, pp. 14–30.
2. Novotny, E., U.S. Patent 1,377,502, May 10, 1921.
3. Bassist, E., U.S. Patent 1,525,531, Feb. 10, 1925.
4. Ducas, C. , U.S. Patent 1,563,731, Dec. 1, 1925.
5. Seymour, F., U.S. Patent 1,647,474, Nov. 1, 1927.
6. Franz, E., U.S. Patent 2,014,524, Sept. 17, 1935.
7. USPQ 137, p. 725, #11421. Technograph Printed Circuits, Ltd. v. Bendix Corp., May 27, 1963.

2

Materials

James Munson

Sheldahl, Inc.
Northfield, MN

2.1 INTRODUCTION

Flexible printed circuitry (FPC) is typically a composite, less than .01 inches thick, of metal foil conductors, and a flexible dielectric substrate. Conductors are usually bonded to the dielectric with an adhesive, but adhesiveless clads have recently become available. Conductors are often protected by conductive and insulative coatings.

Choice of materials for flexible circuits depends on the following:

- Amount of flexure during assembly and operation,
- Electrical requirements of the application,
- Connections to components and other circuitry, and
- Method used for component assembly.

2.2 DIELECTRIC SUBSTRATES

The substrate insulates the conductors from each other and provides much of the circuit's mechanical strength. Use of a flexible rather than a rigid dielectric is the main characteristic that distinguishes flexible printed circuitry from rigid printed circuit boards. This seemingly small difference has immense ramifications in terms of material requirements, circuit manufacturing, and end use applications and characteristics.

In contrast to rigid circuit boards that must be processed in sheet form, the flexible dielectric permits continuous, roll-to-roll handling during processing. Flexible dielectrics can be formed by casting or extrusion, either alone or on the surface of metal foil. Roll-to-roll handling can be used in virtually all flexible circuit processes, such as imaging, etching, plating, die cutting, and component assembly.

Flexible dielectrics expand or shrink during processing to a greater extent than rigid circuit boards. Because flexible dielectrics are generally made in continuous webs, they may shrink in the machine (web) direction, up to .002 inches per inch, and expand slightly in the transverse direction during subsequent manufacturing steps. These effects are seldom a problem if tolerances are made proportional to dimensions, and the capability of FPC to bend when compressed, is used. Plastic films, synthetic papers, and resin-impregnated fabrics have been used as dielectrics in FPC, but polyimide and polyester films satisfy the widest spectrum of requirements. Properties of common FPC dielectrics are compared qualitatively in Table 2.1 while detailed information is provided in Table 2.2.

2.2.1 Polyimide

Polyimide films are produced from a condensation polymer of an aromatic dianhydride and an aromatic diamine. The film is cast in thicknesses of 7.5, 12.5, 25, 50, 75, and 125 microns (0.3, 0.5, 1, 2, 3, and 5 mils), and is processed in widths up to 26 inches. Because these are thermoset polymers, they do not exhibit a softening or melt point. Unlike many thermosetting plastics, polyimide films are very flexible. They have good flexibility and electrical properties across a very wide temperature range, but fairly low tear propagation resistance. They resist soldering conditions and permit repeated resoldering without loss of electrical properties.

Polyimide can be somewhat hygroscopic. Earlier types absorb up to 3% water

TABLE 2.1   Comparison of Dielectric Materials

Property	Polyester	Polyimide	Fluorocarbon	Aramid	Composite
Tensile Strength:	excellent	excellent	fair	good	highest
Flexibility:	excellent	excellent	excellent	good	fair/good
Dimensional Stability:	fair/good	good	fair	good	excellent
Dielectric Strength:	good	good	excellent	very good	good
Solderability:	poor	excellent	fair	excellent	excellent
Operating Temp.					
Continuous:	105–185°C	105°C	+220°C	150–180°C	220°C
Coeff. of Thermal Expansion:	low	low	high	moderate	low
Chemical Resistance:	good	good[1]	excellent	very good	fair
Moisture Absorption:	very low	high	very low	very high	low
Cost:	low	high	high	moderate	moderate

[1]Most are attacked by very strong base.

TABLE 2.2   Properties of Common Dielectrics

	Polyester	Polyimide	Aramid	Glass-Epoxy	Fluoro-carbon
PHYSICAL					
Tensile Strength (PSI):	20–35 K	25–30 K	11 K	35–100 K	4–7 K
% Elongation at Break:	60–165	60–80	7–10	3–5	200–600
Modulus 10^5 psi:	5	4.3	7.9	20–30	0.5
Tear Strength initiation lb/in $\times$ 1000:	1–1.5	1.0	NA	NA	0.6
Tear Strength propagation g/mil:	12–25	8–10	50–90	NA	125
CHEMICAL RESISTANCE					
strong base:	poor	poor	good	fair	good
strong acid:	good	good	good	good	good
oil/grease:	good	good	good	good	good
organics:	good	good	fair	good	good
water abs.:	0.3%	2.9%	8–9%	.05–3%	.01%
THERMAL					
Service Temp. Range, °C:	−60/+105	−200/+300	−55/+200	−55/+150	−200/+200
Coeff. Thermal Expansion, CTE, ppm:	27	20	22	10–12	80–100
ELECTRICAL					
Dielectric Constant;					
1KHZ:	3.1	3.	2.0	4.2–5.3	2.0–2.5
1MHZ:	3.0	3.4	4.5–5.3	2.0–2.1	
1GHZ:	2.8	3	4.5–5.3	2.0–2.05	
Dielectric Strength; volt/mil:	3400	3600	500	240	2000
Volume Resistivity; ohm-cm:	10^{18}	10^{18}	10^{16}	10^{15}	10^{19}

by weight and must be dried immediately before processing or soldering to avoid blistering and delamination. Second generation polyimides, including co-polymers, have lower moisture absorption and a lower rate of thermal expansion. Polyimides have been developed with thermal expansion rates very similar to metals and ceramics. New developments are expected to continue in the polyimide area since extensive R&D efforts are now underway, particularly in Japan.

The most commonly used polyimide for flexible circuits is Dupont's Kapton® film. Similar films are available from several other producers. Some products are cast directly onto metal foil to obtain an "adhesiveless" laminate, as discussed below. Polyimides are one of the more expensive FPC substrates. Table 2.3 lists characteristics of the majority of polyimide films now available. However, with rapid introduction of new materials, the reader is advised to check with circuit manufacturers and film suppliers for updates.[1]

2.2.2 Polyester

This film, used in flexible circuitry, is manufactured from polyethylene terephthalate (PET). For FPC, thicknesses of 25–125 microns (1–5 mils) are the most commonly used. Polyesters have excellent flexibility, electrical characteristics, and chemical and moisture resistance, relatively good thermal and dimensional stability, but low tear resistance. The film is relatively weak under peeling stress when applied perpendicular to the surface. It is capable of intermittent operation up to 150°C. If soldering is required, special provisions, like masking and heat sinking, are necessary to avoid distortion and conductor delamination. Dimensional stability during processing is lower than for polyimide film, but can be improved by preshrinking before processing begins. Polyester film can be adhesively laminated. Other polyesters based on more stable precursors are also available. Polyethylene naphthalate (PEN), for example, has a modestly higher service temperature. None of these other polyesters has offered enough performance advantage to make any progress yet.

Overall, polyester is a nearly ideal flexible base film for flexible circuitry. If it were not for the temperature extremes of soldering, polyester would be used for nearly all of the flex applications. Electrical, chemical, and moisture sensitivity are exceptionally good as can be seen in Table 2-2. The film can be drilled, punched, embossed, thermoformed, laser machined, metallized, dyed, and coated easily. Cost is very low, about 1/20th as much as polyimide. If the drastic assault of the pollution-prone process, called soldering, is replaced with something more moderate, polyester may become the dominant material. Recent advances in solderless assembly, using lower temperature processes, such as thermoset conductive adhesive bonding, allow polyester to be used for electronic assemblies. See Chapter 11 on Polymer Thick Film.

TABLE 2.3   Polyimide Films

Trade Name	Supplier	TCE ppm/°C	% H_2O abs Immersion	Dielectric Const.
Kapton®	DuPont	20 ppm	2.9%	3.4
Apical®	Allied Kanika	21 ppm	2.9%	3.4
Apical NPI	Allied Kanika	8 ppm	2.1%	3.4

2.2.3 Aramid Paper

Aramid paper is a thermoset material based on polyamide—a high performance nylon. The most common material is Nomex® from DuPont, the same product used to make fireman's suits. Rumor has it the name stands for *no more excuses*. Copper laminates, made from nonwoven aramid paper can be readily soldered, but the material suffers from two significant deficiencies: high moisture absorption and poor dimensional tolerance. The product's nonwoven construction makes it difficult to control thickness, leading to a less precisely dimensioned final laminate. Moisture absorption up to 8-9%, is a serious problem because it produces significant dimensional changes and requires thorough drying before soldering. The highly porous nature of this nonwoven construction reduces the kinds of processes that can be run on it. Wet chemical processes like plating, must be avoided because the chemicals are trapped in the paper type structure. Aramid, although moderate in cost, is very limited in the number of its applications.

2.2.4 Other High Temperature Films

Other thermoplastic and thermoset films have occasionally been used for FPC, where their special properties justify the relatively high cost. These include polyetherimide (PEI), polyetherether ketone (PEEK), polysulfone (PS), polyphenylene sulfone (PPS), and polyethersulphone (PES) films. Although some of these polymers can be soldered with great care, they basically offer properties closer to those of polyester with a price nearer to polyimide's.

2.2.5 Reinforced Composite Dielectrics

Composites used in FPC are flexible resins reinforced with high strength fibers or fabrics, and are always made with copper laminated to one or both sides. Their primary advantages are excellent dimensional stability, high temperature resistance, low moisture absorption, and low flammability. Two basic categories of these composites are available, products intended to replace standard laminates and those designed to be shaped into nonplaner forms. A glass-polyester laminate with a relatively soft hand, is sold under the trade name Cosmoflex®. The other type of materials, such as Roger's Bendflex® and Sheldahl's NEL, can be permanently shaped after being fabricated into circuits. The Bendflex® product line has been optimized for shaping and represents a useful concept that allows efficient manufacturing as a flexible product, straightforward assembly as a hardboard, and shaping into a 3-D assembly like a molded board.

Reinforced dielectrics can be machine or hand soldered repeatedly with little adverse effect on the conductor-substrate bond. Reinforced dielectrics are easily processed, using conventional methods for resist coating, etching, plating, and die punching. Circuit laminates with reinforced substrates are much less flexible, have lower tear initiation strength, and are relatively expensive compared to unreinforced substrates.

2.2.6 Fluorocarbon films

These materials have excellent electrical properties, which are constant over a wide range of environmental conditions. They have low dielectric constants and high dielectric strengths that change little with temperature and frequency, making them one of the best materials for high speed circuitry in computers and microwave equipment.

Fluorocarbons have very high thermal capabilities but very poor dimensional stability at high temperatures. They have outstanding chemical resistance, good flexibility, low moisture absorption, and are self-extinguishing.

2.2.7 Miscellaneous

Other films can also be considered, but unless there is an important reason, the standard polymers should be specified. Probably most commonly available plastics films have been tested for application as a flexible base film by vendors and fabricators. Over 2000 plastic films are available so it is possible that some useful polymer has been missed.

Polycarbonate (PC) has been used to some limited extent as a Polymer Thick Film (PTF) base because of its good optical characteristics. This is a useful feature for graphic overlays that have circuitry on the reverse side. The film, however, offers no real advantage over polyester, and it has not been widely embraced as a copper laminate.

Polyvinyl chloride (PVC), a low cost and lowly flammable plastic, has been used to a very limited degree for flex. PVC is a common wire coating so it is not surprising that it has been tried for printed wiring. The real drawback is its poor thermomechanical stability: soldering is difficult at best. Ironically, the material has been used as a kind of conformal coating for flex assemblies because of its high shrinkage factor. A ''shrink wrap'' cover layer is peripherally bond over the component area and then caused to shrink by application of heat. The result looks like an SMT bubble pack. The SMT bubble pack concept is used on some PTF calculators to hold the weakly bonded components to the circuit as well as to protect the circuit.

2.2.8 Polyethylene-Coated Polyester

Poly-coated film and similar hot melt-coated base films are still used today to produce laminates for a special mechanical circuit manufacturing process. Copper foil is weakly bonded to the hot melt-coated base film. The circuit pattern is created by a ''kiss'' cut blanking process that simultaneously heat bonds the conductor trace copper by means of heating elements in the die base. The unwanted copper foil, which is not heat sealed, is peeled away in this efficient continuous roll process. This simple and efficient no etch circuit process has been used to make millions of instrument cluster circuits for the past several decades. Density demands have pushed the process to its line and space width limit, and its use began to decline in the late 1980's. The laminate and the process for making the circuits still represent the lowest cost and most environmentally friendly copper circuit technology yet devised.

2.3 CONDUCTORS

Material considerations for FPC conductors are similar to those of rigid circuit boards. The conductor material must survive processing and provide adequate electrical performance in the service environment. The list of conductors includes elemental metal foils and metal mixtures including stainless steel, beryllium-copper, phosphor-bronze, copper-nickel and nickel-chromium resistance alloys. Both silver and carbon PTF inks are also used.

Conductor properties influence the flexural fatigue life of a FPC assembly.[2] In many ''static'' applications, bending is limited to installation and servicing. In ''dynamic'' applications, the assembly is flexed or folded repeatedly during normal use. For dynamic applications, conductors should be of the minimum acceptable thickness and have a high fatigue ductility. Table 2.4 compares metal properties while Table 2.5 provides guidelines for applications.

2.3.1 Copper Foil

Conductors made of copper foil provide the best balance between conductivity, ease of processing, and cost. The copper foil layer may be electrodeposited

TABLE 2.4   Metal Conductor Properties

Property	Aluminum	Copper	Gold	Iron	Nickel	Silver
Resistivity ohm-cm $\times 10^6$:	2.8	1.7	2.4	10.0	6.8	1.7
Density oz/ft^2 @ 1 mil:	0.22	0.74	1.6	0.64	0.74	0.87
Harness Brinell:	15	42	28	80	110	95
Thermal Conductivity Cal/Sec/cu cm/°C:	0.48	0.92	0.70	0.16	0.14	0.97
Coeff. of Thermal Expan. (TCE) ppm/°F $\times 10^{-5}$:	1.3	0.93	0.79	0.51	0.76	1.05

TABLE 2.5   Conductor Applications

Conductor	Application	Rationale
Copper	95% of all flex circuits	Best balance of properties
Aluminum	Shielding for membrane switch and some circuits	Low cost, but adequate
Silver	Electrical contacts	High conductivity, oxide is conductive
Nickel	Low heat resistance circuits or components	Easily welded
Gold	Conductor and contact plating	Maintains very low plating electrical reistance
Stainless Steel	Resistance heaters, high stress applications	High strength, corrosion resistant
Phosphor Bronze	Corrosion resistant contacts, integrated springs	High corrosion resistance, good elasticity
Beryllium-Copper alloys	Springs	Good electrical, durable spring
Copper-Nickel	Corrosion resistant circuits or heaters	High corrosion resistance, lower conductivity
Nickel-Chromium	High resistance circuits	Low conductivity
Polymer Thick Film	Low cost switches and circuits	Simplified additive processing

(ED) or rolled and annealed (RA). ED and RA copper have different mechanical characteristics that affect their bondability and resistance to bending. ED copper foil is plated from solution onto a rotating drum from which it is continuously stripped. In some adhesiveless FPC materials, copper is first vapor deposited on a film substrate, then electroplated up to the desired thickness.

Rolled annealed copper foil is formed by heating and mechanically rolling ingots of pure copper to the desired thickness. ED copper has a columnar grain structure with grain boundaries perpendicular to the plane of the foil. The grains in rolled copper are like overlapping plates, with boundaries aligned parallel to the foil plane.

The main differences between ED and RA foils are their mechanical properties. Tensile deformation during flexing separates the grains and reduces the flexural endurance of ED foil and its resistance to cracking when folded, compared to RA copper. The platelike grain structure in RA copper provides a higher tensile strength and a much higher resistance to repeated bending, within its elastic limit, than ED copper. It is preferred for FPC's, which are continuously flexed in service.

The columnar grain structure of ED foil results in slightly lower conductivity but more uniform etching and soldering characteristics than RA foil. The ED surface, which is formed against the drum, is smooth and well-suited for photoresist processing. The opposite (solution) side has a rough, granular surface ideal for adhesive bonding to substrates. RA foil is smooth on both sides, which decreases the adhesion of platings and makes it more difficult to solder. For plating and adhesive laminating, rolled copper requires a surface treatment.

Foil surfaces are usually chemically oxidized to increase adhesion, to reduce resist undercutting by etchants, and to reduce bond degradation by plating chemicals. ED foil is easily treated, but treating RA copper is difficult and expensive. A thin layer of zinc is sometimes applied to the surface to increase bond strength and reduce corrosion. Proprietary, stain-proofing treatments are also used.[3]

Electrodeposited copper foil generally contains more pin holes and foreign inclusions than RA copper. Only the highest, printed circuit quality ED copper should be used to avoid excessive foil defects. ED foil is available in both annealed and unannealed grades, as well as a special high-ductility grade having increased break elongation and improved flex life, adequate for many intermediate flex life requirements. Rolled copper is available in a low temperature annealed (LTA) grade, which is harder and easier to handle during lamination to the substrate than ordinary soft rolled copper. The LTA foil is annealed after lamination.

Properties of electroplated and rolled, annealed copper are compared in Table 2.6.

Both ED and RA copper are sold by weight. A weight of 1 oz/sq ft corresponds to a thickness of 0.00135 inches, or 34 microns. Standard weights of copper used in FPC are 0.5, 0.75, 1, and 2 oz/sq ft.

Use of foils less than 0.5 oz/sq ft is increasing, to reduce weight, to improve flexibility, and to improve yields in narrow conductor (< 0.005 inch) circuitry. Thinner foils reduce etching time and conductor undercutting. Ultrathin ED foils are available in weights of 0.375, 0.25, and 0.125 oz/sq ft. The latter two

TABLE 2.6   Properties of Copper Foils
(1 oz. cu, 1.4 mil thick)

Property	Electrodeposited (ED)	Rolled/Annealed (RA)
Purity:	99.8%	99.9%
Electrical Resist. Ohm-cm:	1.8×10^{-6}	1.7×10^{-6}
Elongation at Break:	10%	10%
Fatigue Ductility:	10–25%	150%
Bend Cycles to Failure:	10–100	$> 10^6$

are supplied on a disposable, metal carrier sheet, removed after the foil is laminated to a substrate.

Copper thickness on adhesiveless laminates, plated directly onto a substrate, can range from less than 0.01 oz/sq ft., to over 1 oz/sq ft, and the cost varies directly with thickness. For ED foil thinner than 0.4 oz/sq ft, the cost of the copper varies inversely with thickness, because of handling difficulties, surface defects, and the carrier sheets. In gauges from 0.4 to 1 oz/sq ft, untreated ED foil is similar in cost to RA. Above 1 oz/sq ft, ED is more expensive than RA because more energy is needed to plate-up equivalent thicknesses. The cost of RA foil also varies inversely with thickness. Surface treatment adds a negligible amount to the cost of ED, but nearly doubles the cost for RA. Table 2.7 recommends the best type of copper for various applications.

2.3.2 Other Metal Foils

A few other metals are occasionally used in FPC for special purposes. Aluminum is used for electrical shielding or to replace copper in very low-cost circuits. It cannot be soldered or welded with conventional assembly equipment, but electrical connections can be made with conductive adhesives. Phosphor bronze and beryllium copper foil provide conductors with integral leaf springs and corrosion-resistant contacts, where precious metal plating is not required. Soft ferromagnetic foils are used for magnetic shielding. Stainless steel, Monel, Inconel and other alloys are used for thin, flexible resistance heaters and circuitry requiring high strength and corrosion resistance.

2.3.3 Polymer Thick Film (PTF)

PTF conductors consist of finely divided conductive material, like silver or carbon, in a polymer binder, like polyester, epoxy, acrylic, urethane, or vinyl. Conductor patterns are typically made by screen printing PTF inks onto a di-

TABLE 2.7   Selection of Copper Conductors for FPC

Application	Recommended Type
Dynamic or continuous motion:	RA
Very fine line etching:	RA
Non-dynamic cables, but subject to vibration:	RA
Prototypes, short runs:	RA
Double-sided plated through hole circuits:	RA, also ED
Large radius flexing with modest number of cycles:	Annealed ED
Nondynamic:	ED
Bend to install (> 100 mil radius):	ED

electric substrate and during the coating. PTF printed circuitry was originally used for low-cost, low-power circuitry, such as membrane keyboards or for jumpers on copper foil FPC. PTF circuitry is much lower in cost than copper, because conductors are formed in a single step without resist placement, etching, stripping, and cleaning. PTF coatings may be printed-through holes in thin insulating substrates to join several layers of conductors. The flex life of PTF coatings is similar to that of copper foil of equal thickness. Reduction in conductivity caused by repeated flexure of PTF conductors may be partially recovered by creep in the resin, when the material is at rest.

The bulk resistivity of silver filled PTF ranges from 75 down to about 10 mohm per square per mil of thickness—at best an order of magnitude greater than for copper foil conductors. Typical PTF conductors are 0.4 to 0.7 mils thick. On thin polymer dielectrics, in free air, these can dissipate 2 W of continuous power per sq in. without thermal or chemical breakdown and up to 60 W/sq in. when bonded to a heat sink. PTF conductors are difficult to solder by conventional means. Their connections are usually made by pressure contact or with conductive adhesives.

The introduction of PTF circuits having components assembled with conductive adhesives, may increase the use of these materials in applications formerly made exclusively with copper and solder. Considerable development effort is going into this area because of cost savings and the intrinsically low pollution associated with the circuit and assembly technologies.

2.4 ADHESIVES

Most FPC is made with adhesives even though the substrate and conductor layers may be joined directly, by forming one or the other in place. Adhesives must be compatible with both substrate and conductor layers, and withstand the conditions used in FPC manufacture without delamination, excessive flow, or degradation of properties. Depending on the degree of chemical crosslinking, all FPC adhesives in common use have good chemical resistance and fair to excellent flexibility. Most can be made flame retardant using additives.

FPC adhesives are generally coated onto the dielectric, which is then laminated to the conductor foil, either in a platen press, or by passing the composite between heated rollers. Post curing of the adhesive at elevated temperature is often required after roller lamination. The adhesives most widely used are polyesters, modified epoxies, and acrylics. Polyimide adhesives are gaining favor, especially for multilayer and rigid-flex constructions. See Table 2.8.

2.4.1 Polyester

Polyester adhesives are low-temperature thermoplastics. They can be modified by partial crosslinking. Modified, or catalyzed, polyesters have high tempera-

TABLE 2.8   Properties of Flexible Circuit Adhesives

PROPERTY	Polyimide	Polyester	Acrylic	Mod.-Epoxy
Peel Strength Lb/in:	2.0–5.5	3–5	8–12	5–7
After Soldering:	no change	?	1–1.5 × higher	variable
Low Temp. Flex:		all pass IPC-650 2.8.18 @ 5+		
Adhesive Flow:	<1 mil	10 mils	5 mils	5 mils
Temp. Coeff. of Expan.:	<50 ppm	100–200	350–450	100–200
Moisture Absorption:	1–2.5%	1–2%	4–6%	4–5%
Chemical Resistance:	good	fair	good	fair
Dielectric Constant @ 100KHZ:	3.5–4.5	4.0–4.6	3.0–4.0	4.0
Dielectric Strength; Kvolts/mil:	2–3	1–1.5	1–3.2	0.5–1.0

ture properties nearly equal to those of thermoset adhesives while retaining the flexibility of thermoplastics. Modified polyester adhesives are used because of their flexibility and ease of processing. They can be used in applications requiring both machine and hand soldering.

2.4.2 Acrylic

Acrylic adhesives are thermosetting materials. They have a higher resistance to soldering conditions than polyesters and modified epoxies. Flexible laminates with acrylics are generally available only in sheet form, which can significantly increase the cost of FPC made with these adhesives. Environmental concerns have spurred the search for waterborne, acrylic resins. Success in these developments will permit roll-to-roll manufacture and reduce the cost of this material for FPC in the future.

2.4.3 Modified Epoxies

This type is the most widely used adhesives for rigid circuit boards because of their resistance to high temperatures, both for soldering and in service. Pure epoxy adhesives are less flexible than polyesters and acrylics, and must be modified for FPC by the addition of other polymers, like polyesters, to increase flexibility. These epoxies are thermosetting materials.

2.4.4 Phenolic

This class of adhesive is popular for tape automated bonding (TAB) materials. They have physical and thermosetting properties similar to epoxies. Because of the curing process, they are usually available only in single-sided laminates. Their flexibility can be enhanced by additives, but other adhesives are better for dynamic bending applications. Phenolics do not adhere as well to polyimide film as polyester or epoxy adhesives.

2.4.5 Polyimide

Polyimides can be formulated to withstand temperatures as high as 700°F. They evolve water on curing and require very stringently controlled post-curing processes. Polyimides have relatively low bond strength to polyimide film and are less flexible than acrylics. Considerable R&D is going into this class of adhesives, primarily in Japan, so major improvements can be expected. A primary attribute is very low TCE value—an important characteristic for multilayer circuits.

2.4.6 Fluorocarbon

Fluoropolymer-based adhesives provide good electrical and environmental performance over a broad range of conditions. Fluorocarbons have fairly low melting temperatures, are heat sealable, but are dimensionally unstable during soldering. Table 2.9 provides a qualitative overview of the above adhesives.

2.5 ADHESIVELESS LAMINATES

Originally, films and foils for FPC were always bonded together with a third material. These adhesives can significantly affect the reliability of plated-through holes, both during drilling and plating operations, and during thermal, Z-axis expansion in service, which is important in multilayer applications. Dielectric properties of adhesives may degrade performance of high speed circuitry. Printed circuit resistance to thermal and chemical stress is often governed by the adhesives used.

TABLE 2.9   Comparison of Adhesives

PROPERTY	Polyester	Acrylic	Mod.-Epoxy	Polyimide	Fluoro	Phenolic
Temperature		very				
Resistance:	fair good	good	excell.	good	good	good
Chemical		very				
Resistance:	good	good	fair	good	excell.	good
Electrical:	excell.	v.good	good	good	good	good
Adhesion:	excell.	excell.	excell.	good	good	good
Flexibility:	excell.	good	fair	fair	excell.	good
Moisture						
Absorption:	fair	poor	good	fair	excell.	fair
Cost:	low	moderate	high	very high	high	moderate

To avoid these problems, FPC composites can be produced without an adhesive, either by coating the dielectric with a conductive material, or the converse. A thin coating of copper may be applied to dielectrics by vacuum deposition, by sputtering, or by electroless plating, and built up by electroplating. Polyimide resin may be coated onto copper foil and cured.

Large development programs to produce adhesiveless products are underway in Europe, Japan, and the United States. New products began their serious introduction in the mid-1980's. Copper-polyimide clads are now offered with thicknesses ranging from 0.3 microns (about 10 millions of an inch) to about 35 microns (1.4 mils). We predict that adhesiveless products will make strong inroads into the flex circuit materials market. The products and the circuit processes that they make possible, have some intrinsic advantages over adhesive laminates.

Adhesiveless copper-polyimide can be made in thinner, more easily etched copper weights. The old standard of 1.4 mils (1 oz/sq ft), really a defacto from the copper mills, is too thick for most modern applications where current demand is down and density requirements are up. The extra copper thickness requires longer etching, produces more waste, and gives poorer pattern resolution. Adhesiveless product, produced in 1/4 oz or even 1/8th oz thickness, is much more suitable for fine line circuitry. The negative factor is that the copper is mostly electrodeposited (ED) foil, an unavoidable by-product of the adhesiveless technology. This means that the product is not suitable for dynamic flexing applications. Improved copper plating technology may improve the ED copper. Low Temperature Annealable (LTA) electroplated copper is already finding applications in less demanding flexing applications, such as CD players. The LTA copper must be brought to a relatively high temperature to produce the annealing. This can require an extra step unless there is a cover layer or multilayer lamination step where a "free" heat cycle is achieved. Although there are adhesiveless processes that start with RA copper, they have the same higher copper thickness requirements as those for conventional laminates.

Adhesiveless products can allow semi-additive processing. Basically, the circuit patterning method involves placing a plating mask over very thin copper coated on a base dielectric film. Copper is then plated-up in the exposed areas to the desired thickness. Conductor sidewalls can be very straight because no etch factor is involved. The plating mask or resist is stripped off, and the entire circuit undergoes flash etching, which removes the thin background copper and an insignificant amount of the plated-up traces.

One advantage of adhesiveless copper-polyimide is the ability to laser machine and chemically mill the dielectric. Most, but not all polyimides, can be etched with a strong chemical base to create "windows" and holes. This windowing feature is of great value for producing Tape Automated Bonding (TAB) features in a circuit. Virtually all polyimides respond well to excimer laser machining. The absence of an adhesive permits clean fabrication of micro (below .001 in.) and TAB windows. These adhesiveless clads, therefore can be used to make 2-layer TAB and Integrated Beam Lead (IBL) within a flex circuit. These topics are covered in Chapters 8 and 9.

2.6 CONDUCTIVE COATINGS

2.6.1 Platings

Bare copper readily oxidizes or tarnishes, if not protected from the environment. Chemical plating is the most common process for covering copper. Plating materials include solders nickel, gold, and tin. Plating can provide a variety of features including corrosion protection, wear resistance, lubricity, increased, or reduced hardness, solder for bonding and reduced electrical contact resistance.[3] Table 2.10 indicates the general applications for the common plating metals. Table 2.11 describes common application methods for the standard metal platings.

Metallic plating is used on flexible circuits to:

- Provide interconnections between circuit layers
- Increase current carrying capacity, or mechanical strength
- Protect exposed conductors from corrosive environments
- Assure solderability during component assembly
- Serve as an etching resist

Some circuits are selectively plated to permit later electrical connection or mechanical bonding of components. Plating copper onto the walls of holes in the circuit substrate is the most common technique for interconnecting several foil layers in a multilayer circuit. Solder plating is often a cost-effective way of protecting exposed copper conductors from environmental extremes and for assuring solderability. When circuits must remain solderable for only a short time,

TABLE 2.10   Plating Materials

Metal	Applications
Copper:	Most commonly used plating for circuit fabrication. Used to create Plated-Through Holes.
Solder:	Most common protection for copper. Provides a good solderable surface. Can be used as etch resist.
Nickel:	Base plating prior to gold plating to create a barrier.
Silver:	Used for contacts. Rarely used now. Can be applied as PTF ink.
Gold:	Excellent corrosion protection. Very durable, low ohmic surface. Not a good surface for soldering.
Tin:	Used for thermocompression bonding surfaces.

during staging or inventory holds, the most economical coating is a flux lacquer or oxide inhibitor. These have a shorter shelf life than solder, but can be effective up to several months if circuits are sealed in plastic bags and stored in a cool, dry environment. Table 2.11 summarizes various plating methods. These methods are discussed in detail below.

Electroplating.   Metal deposition is done on the conductors by electrolysis, using electric current as the driving force. Copper, solder, gold, nickel, and tin are commonly electroplated on FPC.

Electroless Plating.   A metal deposition method in which a chemical oxidation-reduction reaction is used to reduce metal ions to metal on a nonconductive surface. Nickel, gold, and copper electroless coatings are available.

Solder Coating (Tinning).   Deposition method on a metal surface by direct contact with molten solder.

Screen Printing.   This process involves printing a paste of powdered solder and flux through a screen or stencil, followed by fusion of the solder with IR radiation or hot gas.

2.6.4 Conductive PTF Coatings

Carbon or graphite filled polymer thick films (PTF) have been used to protect conductor surfaces from corrosion and oxidation and to inhibit metal migration, especially of silver PTF conductors. These are particularly suited for pressure

TABLE 2.11   Solder Coating Techniques

Plating Method	Characteristics	Disadvantages
Electroplating	Thickness: .001–.0001″ Can control thickness to ±.00002″. Very smooth, uniform. Moderate cost.	Reduced yields. Requires treated copper. Conductors must be bussed together.
Electroless	No bussing required.	Slow process, limited to very thin coatings.
Solder Coat (Tinning)	Fast. Lower cost, good yields, better solderability. Bussing not required. Does not require treated copper.	Large thickness variation. Nonuniform surface. Not always 100% coverage. May need touch up.
Screen Printing	Typical thickness .0015″. Allows selective coating. Good thickness control.	More expensive.

contacts, such as membrane switches and connections to components, such as liquid crystal displays (LCD). PTF coatings are low in cost compared to plating methods. A typical 0.0005 in. thick carbon coating adds about 1 to 10 milliohms to a pressure contact. PTF coatings are selectively applied by screen printing. Carbon ink may be applied to bare copper or to solder plated copper. In the latter case, the carbon serves as a softer, more flowable contact surface for devices, such as LCD's.

2.6.5 Chemical Treatments

It is often desirable to maintain a bare copper surface for assembly considerations. Solder plate can melt and flow in all areas during reflow assembly processing. Reflowing of solder under the solder mask is undescribable because the movement of the solder can degrade the mask. Therefore, processes have been developed to protect copper without using solder while still maintaining solderability. This permits the solder mask to be applied directly over the copper.

Two classes of treatments are in use today, those which chemically react with copper surfaces and others which form a thin, protective, but removable coating. The chemically reactive materials are usually nitrogen base chemical compounds, which form molecular complexes with metals like copper. Benzotriazole and imidazole compounds are the common active ingredients in proprietary treatments. They form invisible, submicron thick deposits that are decomposed at soldering temperatures to leave a purer copper surface. These molecular-thin coatings do not interfere with electrical contact because a very slight abrasion removes the complex. Treatments can last 6–12 months if circuits are packaged. A 30-day use safety factor is more advisable. Modified triazoles and imidazoles can provide longer protection, but removal can be more difficult.

The next class of protectant is limited to thin rosin coats. Such a film will protect the copper surface and often aid in soldering. However, aging or excessive heating prior to soldering can polymerize the rosin and make it interfere with soldering. Coating thickness is much more variable than the reactive treatment materials. These rosin coatings, under such trade names as Entek®, should be specified only if tested and found to be compatible with the assembly process and the end use application.

2.7 NONCONDUCTIVE COATINGS

Protective films or coatings may be selectively applied to the surface of FPC to protect it from moisture, contamination and abrasion, and to reduce conductor stress during bending.

2.7.1 Cover Layers

These coatings consist of an insulating film coated with an adhesive. Pad access holes and registration holes are drilled, punched, or laser machined into the film. The cover layer is registered over the etched conductor pattern and laminated under heat and pressure.

To reduce conductor damage from frequent bending, the thickness of the cover layer should be about the same as the thickness of the dielectric layer. This locates conductors near the neutral surface and reduces conductor stress during flexing.

The same film material is often used for both the dielectric substrate and the cover layer of a flexible circuit. The most commonly used materials are polyester film coated with polyester adhesive, polyimide film with acrylic adhesive, and polyimide film with epoxy adhesive.

2.7.2 Cover Coatings

Cover coatings are formed from solution or from liquid polymers. The material is screen printed onto the circuit leaving pad areas exposed. The polymer resin is then cured either thermally or by UV radiation forming a permanent, thin, tough coating. Acrylated epoxy, acrylated polyurethane, and thiolenes are used for flexible circuit cover coats. These are liquid polymers that require no solvents during coating. They are cured using UV radiation.

Electrically, insulating elastomers, such as silicone, polyurethane, and butyl rubber have not been widely used on flexible circuits because they stretch and deform in service. Polyester and acrylic resins are sometimes used as liquid cover coats on polyester circuits.

2.7.3 Photoimagable Solder Masks

These materials are coatings in which conductor access holes are formed by photoprocessing. Both precast films and liquid masks are available. The films are heat laminated over the etched FPC. Liquid forms are applied with a roller, curtain coater, or similar method.[4] The masks are exposed to light through a photographic negative that creates access holes and other unprotected areas. The coating cures in the photoexposed areas. Uncured material is chemically stripped, leaving a patterned covering.

2.8 FLEXIBLE ELECTRICAL LAMINATES

A flexible electrical laminate is a composite of a metal conductor bonded to a dielectric substrate. Flexible electrical laminates are the base materials for FPC (see Table 2.12). These laminates offer several advantages to the manufacturers of FPC. These are discussed in further detail below.

2.8.1 Matching of Component Properties

Flexible laminates are formulated using dielectrics and adhesives with physical, electrical, and thermal properties that are complementary. For example, high temperature adhesives are used with polyimide substrates to retain the excellent thermal properties of polyimide. Less flexible adhesives, such as epoxies, are used with reinforced dielectrics, since these substrates have limited flexibility.

2.8.2 Product Performance History

Quality and reliability data have been compiled to document the performance of standard laminates. Circuit manufacturers are assured that laminates will consistently meet their property specifications. Manufacturing data are available to enable the selection of a laminate compatible with FPC fabricators' manufacturing processes or the adjustment of processing parameters to accommodate laminate properties. Studying common applications of standard laminates will help a designer to select appropriate materials for similar applications.

2.8.3 Properties of Flexible Electrical Laminates

Flexible electrical laminates are designed so that they can bend, twist, and roll without fracturing either the metal conductor or the dielectric substrate. The amount of flexibility required is determined both by the end application and by the method of FPC manufacture.

Flexible circuits often encounter more severe conditions during manufacture than in service. The composite of conductor, adhesives, and dielectrics must resist degradation by the following:

- mechanical operations, such as drilling, shearing, and handling;
- process chemicals, such as plating and etching baths, strippers, and cleaners; and

TABLE 2.12   Commonly Available Copper Clad Laminates

Dielectric Film	Adhesive
Polyester	Polyester, FR polyester, modified epoxy
Polyimide	Acrylic, FR acrylic, modified epoxy, polyester, polyimide, phenolic.
Fluorocarbon	Epoxy, acrylic, none (fused), fluro.
Aramid	Acrylic, modified epoxy, phenolic.
Composite	Modified epoxy, polyester, other.

- high temperature conditions, such as laminating and soldering.

An adhesive peel strength of 4 lbs/in. between conductors and dielectrics is usually adequate, if the adhesive has good thermal resistance to soldering and good chemical resistance. Peel strength has little significance to the processability of adhesiveless FPC materials.

The properties of three common types of electrical laminates are compared in Table 2.13. Table 2.14 is a listing of recommended adhesive and dielectric types for various applications.

2.8.4 Selecting Flexible Laminates

Example 1:   An application requiring continuous high-temperature operating conditions and more than 1,000,000 flex cycles.

Laminate: Polyimide film × acrylic adhesive × treated RA copper.
This laminate had the best temperature and solderability properties and adequate flexibility to meet the one-million cycle requirement. Treated RA

TABLE 2.13   Flexible Electrical Laminates

Base Film:	Polyester	Polyimide	Composite
Adhesive:	Polyester	Acrylic	Epoxy
PROPERTY			
Thermal Stability:	fair	excellent	good
Dimensional Stability:	fair	good	excellent
Tear Resistance:	good	fair	excellent
Flexibility:	excellent	very good	fair
Electrical:	excellent	very good	fair
Water Absorption:	excellent	fair	good
Cost:	lowest	highest	moderate

TABLE 2.14   Applications for Flexible Electrical Laminates

Applications	Dielectric Film	Adhesive
Disk drives, telecom, high reliability	Polyimide	Modified epoxy
Military, rigid-flex, appliance	Polyimide	Acrylic
Advanced multilayer, mainframe computer, automotive, high temp.	Polyimide	Polyimide
Consumer electronics with minimum soldering, automotive cluster, printers, telephone	Polyester	Polyester
Telecom PBX, switching.	Composite	Epoxy
Under hood automotive, bus bar	Aramid	Acrylic

copper is required for dynamic flexing. Copper surface treatment will assure solderability.

Example 2:   Same as application in Example 1 except that soldering is not required in circuit manufacture, and a low cost circuit is desired.

Laminate: Polyester film × polyester adhesive × RA copper.
Polyester will probably withstand the operating temperatures during application. Since soldering is not required, this temperature performance is likely to be adequate. Polyester materials will yield a low-cost circuit. RA copper is needed for dynamic flexing.

Example 3:   An application requires better than 1 mil dimensional stability and only requires flexibility in processing or installation.

Laminate: Reinforced substrate × epoxy laminate × ED copper.
This material offers excellent and consistent dimensional stability. It has a flex capability of 25-50 cycles which is more than adequate for most flex circuit processing and installation operations. ED copper is adequate for static flexing applications.

Example 4:   An aerospace interconnection application requiring high reliability and exposure to multiple soldering operations.

Laminate: Polyimide film × fire retardant acrylic adhesive × ED copper.
Polyimide substrate is needed to withstand the soldering operations. Fire retardancy is usually a major concern in aerospace and military applications. ED copper is needed for fine line circuitry.

Example 5:   A flexible-to-rigid circuit board application requiring a low cost circuit.

Laminate: Polyester film × polyester adhesive × ED copper.
Polyesters are the lowest cost materials used in flexible electrical laminates. Unless the circuit will be exposed to very high temperatures, the thermal properties of polyesters should be adequate for this type of application. If flexing in packaging is all that is required, ED copper can be used.

Example 6:   A high-volume multilayer circuit board application requiring excellent dimensional stability, low moisture absorption, and suitability for hand or wave soldering.

Laminate: A laminate using a reinforced composite substrate and treated ED copper.

Polyimide dielectrics cannot be used because of their high moisture absorption. Only reinforced composites will give the dimensional stability required in this application. Treated copper assures solderability.

Example 7:   A high-quality flexible circuit application requiring dynamic flexing, solderability, and excellent resistance to acids, alkalis, and solvents.

Laminate: Polyimide film and acrylic adhesive × treated RA copper.
Soldering operations require the use of a high-temperature material such as polyimide or reinforced composites and treated copper. Reinforced materials are not flexible enough to be used in dynamic flexing applications. Both polyimides and acrylics have good chemical resistance.

References

1. English, L. K. New Directions in High Performance Films. *Materials Engineering*, July 1988.
2. Shepler, T. H. Flexible Interconnections—Materials and Design. *Circuit World*, Vol. 13, No. 3, 1987.
3. Willis, R. Solderable Finishes for Surface Mount Substrates. *PC Fabrication*, Oct. 1989.
4. Tuck, J. An Update on Liquid Photoimageable Solder Masks. *Circuits Manufacturing*, May 1989.

3

Constructions

Joe Fjelstad

J. Fjelstad Associates

and

Ken Gilleo

Poly-Flex Circuits, Inc.

3.1 INTRODUCTION

There are many options available to the designer or packaging engineer charged with the creation of a flexible circuit design for a new product or system. The decision of how to construct a flexible circuit is firmly linked to the requirements of the final product. Failure to take into account those requirements can result in unnecessary expense in manufacturing the product. There is a high probability of over designing the circuit and adding unnecessary expense, and there is also an equal chance of under designing the circuit and having it cause problems in production or fail prematurely in use. A clear understanding of circuit constructions will help avoid these pitfalls.

The highly custom nature of flex has led to many ways of achieving a particular design, unlike hardboard which is much more standardized. The authors will discuss the construction and design options and set forth strategies for their implementation and use. While it is impractical to provide a blueprint for every design situation, numerous exemplary cases will be provided for clarity. The information provided here is purposely generic and will describe the most common form of the subject construction. Many manufacturers of flexible circuits have developed their own special techniques, and many are unique to a specific company. While some processes may be company-specific, they enable reliable and cost effective products to be created. This also means that different flexible circuit manufacturers can have processes that are better suited to specific designs and constructions. For example, a continuous web process may be well suited to high volume, low cost single- or double-sided circuitry, but not fine line.

3.1.1 Construction Strategy

Generally, the simplest scheme for a flex circuit implementation is the best strategy. A general guideline is to start with the simplest constructions first to better grasp the subtleties of flex circuit design and use. This strategy may need to be altered because of the particular design requirements, clever ideas may allow the design to stay simple. For example, we could start a single-sided pattern and find that problems arose because the circuit needed to be connected from both top and bottom. Rather than jump immediately to a double-sided construction, we would first try folding the circuit in a configuration that would bring the conductors to both sides. Should access to both sides be required in the same location, we would then move to a double-access, or back bared construction, where the base film is selectively removed to accommodate an electrical connection. It is clear that the high versatility allows options not available in rigid circuits. This makes it essential that the designer explore all the construction and configuration possibilities because a good design solution is not always obvious with versatile and adaptable flexible circuit technology.

In the following sections of this chapter, you will be introduced to the many different constructions used in fabricating flexible circuits. We will cover simple single-sided one conductor layer flex circuits to complex multilayered constructions that utilize multiple breakout tentacles for interconnecting the circuit on many 3-D planes. We will also provide information on the major manufacturing processes that are among the available choices for fabricating these ingenious circuits and interconnection devices.

3.1.2 Increasing Circuit Density

As a final note of introduction, it is worth taking a few moments to investigate some of flex circuit technology's unique ability to increase circuit packaging density. Flexible circuits have the unique ability to increase circuit density in several ways. The most obvious, but often overlooked concept, is to configure flex into a multiplaner form. The flex circuit can be folded and shaped into a very dense 3-D package. These new packages often occupy a small fraction of the volume of mere conventional design approaches. The flex designer must think volume not just real estate area.

Another density-increasing advantage is achieved by mating flexible circuits with other high density technologies, such as Surface Mount Technology (SMT). There is a multiple synergy achieved by mating flex and SMT.[1] Here the advantage is doubled by providing small components to compliment and enhance flex technology's packaging ability. The advantage is also enhanced from a reliability standpoint by providing the surface mount components a compliant substrate to mitigate the effect of thermal coefficient of expansion mismatches. SMT-Flex is covered in detail in Chapter 9.

Density can be improved by employing standard approaches, such as increasing layer count, reducing feature sizes, or designing with features normally thought to be reserved for rigid board fabrication, such as blind and buried vias. Noting that such approaches frequently come at a fairly high premium, the designer may wish to hold them in reserve until all other approaches have been exhausted.

Yet one more, very powerful approach to density is the integration of features within the flexible circuit. Chapter 8 covers this topic in detail so we will only provide a few examples that deal with density enhancement. One concept, which is only now emerging, is called TAB-FLEX, TAB-Featured Flex, or Integrated Beam Lead Flex. Tape Automated Bonding, or TAB, is a specialized flex circuit designed to serve as an interconnect to an IC. But TAB tape is subsequently attached to a circuit board. Because TAB and flex circuitry are one in the same, we can simply design the TAB feature into the flexible circuit. Such constructions are designed to accept bare IC chips directly onto flex circuit signal traces. The result is true Chip-On-Board (COB) with only one interconnect between the IC and the circuit. This type of product will prove of great value to those seeking the highest levels of circuit density.

3.2 SINGLE-SIDED FLEXIBLE CIRCUITS

Single-sided flexible circuits are the simplest example of the technology. This type of flex has only a single conductor layer in the finished construction. Single-sided flex is also the most broadly specified type of flex circuit, the least expensive, and the flex circuit type produced in the greatest volume. Single-sided flex is readily adapted to efficient continuous web processing, which makes it popular for low cost high volume usage. Single-sided circuits are also specified for dynamic, or continuous flexing, applications, such as printers and disk drives. Following are descriptions of the variations in basic manufacturing processes that can be employed to fabricate single-sided flex circuit constructions.

3.2.1 Conventional Single-Sided Construction

The most basic method of construction for a single-sided flexible circuit is commonly referred to as ''Print & Etch Technology.'' It is also often called ''Subtractive Processing'' because conductor foil is removed to create the conductor pattern. Constructions of this type are commonly manufactured by placing a chemically resistant image of the circuit directly onto the metal surface of the flexible copper clad laminate. Acid or base resistant ink is screen printed and cured by heat or UV radiation. The next step involves etching the unwanted copper, leaving the desired circuit pattern. Exposed areas of copper are "dissolved" away by acid, alkaline, or other reagent, which can convert copper to one of its soluble salts. The final step of the circuit patterning process involves removing the etch resistant coating.

Etching generally occurs at an equal rate for all exposed areas. This means that as the exposed copper is etched down, a beveled edge results at the edge of all the conductor traces. The end result is that the cross section of a conductor will have a particular angle that is associated with the etchant and the process. The rate of etching down vertically to etching horizontally is called the etch factor. All things being equal, the factor is $1:1$, and the resulting bevel angle is $45°$. However, the etching process has been improved to the extent that the etch factor usually exceeds $1:3$. Use of etch spray nozzles, for example, allows more vertical etching to take place. Modern etchants also tend to have a higher vertical rate thereby increasing the etch factor. Copper grain structure can also play a roll, but even with today's highest etch factors, which approach $1:8$, there will still be trace wall beveling.

More precise, finer line circuits can be produced if the etch resist is placed photographically. The photoresist method involves applying a photosensitive film or liquid, exposing it to selective light through an artwork, and developing away the unwanted material. Etching is carried out, and the photomask is removed or stripped. Both image placement processes are covered in more detail in other chapters, such as Manufacturing—Chapter 7.

Most applications require that specific areas of the circuit be protected from the environment, electrical contact or solder. A protective cover layer film can be bonded to the circuit. Alternatively, a cover coat or a solder mask can be screen printed on. The protective covering has selective openings to provide access for component mounting or interconnection. Holes in the cover layer are normally produced by drilling or punching prior to bonding to the circuit. Openings in the cover coat are produced by the screen stencil making this method ideal for quick change and recurring revisions. More will be said about protective coverings later.

3.2.2 Die Cut Process

Most of the world's circuits are made by etching copper in the common subtractive processes just described. Another less common process involves mechanically cutting the circuit traces and peeling away the unwanted background foil. The pattern blanking, or die cutting process, has been practiced for over 25 years as an economical means of making flexible circuitry. It has served the automotive cluster area well, providing a very cost effective means of producing the large, wide conductor circuits used behind the instrument cluster. The process is still used today but is falling into decline as increasing density demands exceed the technical limits of the method. Packard Electric Corporation, in the U.S., has used this process until recently to produce instrument cluster circuits for General Motors cars. The process has been used exclusively for single-sided circuitry, although a double-sided circuit is theoretically possible. However, a compatible side-to-side interconnect process would have to be developed.

The process starts with a lightly bound copper-polyester laminate, held together with a heat activated adhesive. Initially, the laminate has a low peel strength so that unwanted copper can be easily peeled away. The roll laminate is run through a large hydraulic press with a circuit pattern kiss cut die. The die cuts the copper circuit pattern while only slightly indenting the polyester base film. The blanking tool is heated so that the adhesive is activated, strongly bonding the copper to the base film where contacted by the die. The area beyond the conductor traces is not heated and bond strength remains low. The roll of circuits is wound up while the scrap copper is rolled up separately. Because the adhesive coating is on the polyester, the scrap copper is essentially pure and readily recycled.

The process is elegant in simplicity, but crude in terms of density and quality. The method is probably the lowest costing circuit making process ever developed because it amounts to a single step process. Large rolls of laminate are fed into an automatic press, which produces several square feet of circuitry for every stroke. No pollution is involved, except for perhaps the adhesive coating step. This efficient process has several major drawbacks that have limited its use beyond the relatively simple cluster circuit. Tooling is tremendously expensive compared to other methods. Large precision dies cost tens of thousands of dollars. Engineering changes are prohibitive in most cases because of the unalterable nature and cost of the tool. Trace width is limited to around 40 mils for the original type of tooling. Tooling advances, particularly in Japan, have dropped the width limit to less than half, but in an electronics world where size is measured in microns, the value of this kind of density is rapidly waning. The designer should consider this method only for very high volume, wide trace circuits.

3.2.3 Semi-Additive Processing

Semi-additive processing, also sometimes called semi-subtractive processing is similar in many respects to the conventional processing image and etch process outlined previously. The major difference is that the laminate is coated with only a very thin layer of copper to which additional copper is added by electroplating. An imaging mask is applied, and copper is added to the conductor paths by plating. The photomask has a negative format, openings where the pattern is desired. This is just the opposite of the mask used for etching. Figure 3.1 shows a simplified semi-additive process.

There are several advantages to the semi-additive approach. An important one is that a laminating adhesive is eliminated. In most situations, the adhesive is the weakest link especially in regards to thermal and chemical stability. Deposition of metal directly to the base dielectric provides a good foundation for building a superior flexible circuit. The presence of a thin but continuous conductive film means that the circuit copper can be built up by electroplating. The thin copper serves as an electrical bus. On the negative side, electroplating a masked thin copper substrate is a challenge. Differential current density of the masked clad can create unequal plating rates resulting in uneven trace thickness.

The semi-additive circuitry process typically starts with a thin seed layer of copper that covers the entire dielectric base film. The copper, or other platable metal, may be vacuum deposited or flash plated directly onto a sensitized dielectric. Other methods of metal deposition can be used, but vacuum and chemical plating are the popular ones, with vacuum deposition gaining. Flex is an ideal circuit product for the vacuum deposition process because it can be processed in roll form. This means that rolls of base film weighing thousands of pounds, can be run in web type vacuum metallizers. Sputtering, ion plating, and thermal evaporation vacuum deposition methods have been claimed. Several vacuum metallizing companies offer products with anywhere from 0.1 micron to several microns of copper.

3.2.4 Fully Additive

Fully additive flexible circuits can be produced by at least two different techniques—electroless copper plating and polymer thick film. Each of these methods has special merits and applications. Both methods involve applying or depositing the conductive material only onto the desired area. There is no subtractive step.

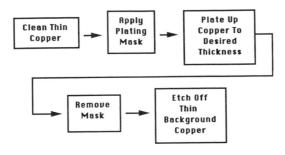

FIGURE 3.1   Semi-additive circuit process diagram

Fully additive technologies offer the distinct advantage of being able to reduce the amount of processing required to create a circuit. They therefore present some of the greatest opportunities for cost reduction among flex circuit constructions. However, as with semi-additive flex circuits, fully additive flex copper circuits are normally restricted to static and limited flex cycle dynamic applications. Again, be aware that advances in technology may make it possible to design fully dynamic flex copper circuits using the additive method at some point in the future. Today, at least one limited motion dynamic flex circuit for a CD player, uses plated copper.

Electroless Copper.   The use of electroless copper to form circuits is predicated on the principle of surface catalysis to allow selective copper deposition. The catalysts used for such processing are normally precious or semi-precious metals. The most commonly used among these is palladium.

A standard manufacturing approach to additive circuits begins with a catalyzed substrate. A plating mask is applied having the desired circuit pattern but with a negative or reversed image, similar to the type used for semi-additive processing. The imaged substrate is then immersed in an electroless copper solution, and plating initiates on the exposed catalytic surfaces. Plating is allowed to continue until the desired amount has been deposited. Various modifications of the method are also known, but the basic process involves selectively catalyzing the electroless plating of metal onto the dielectric.

There are also some ''maskless'' processes that have been developed, but these have not gained commercial significance. They involve coating the substrate with a catalytic salt or one that becomes catalytic with a specific treatment. Another modification of the maskless process involves applying a photosensitive ''anchor'' to the dielectric. Exposure to light deactivates the material, allowing the catalyst to be precipitated only by the unexposed material. A process developed several years ago by AT&T[2] starts with polyimide film treated with tin salt in a lower oxidation state. Exposure to light oxidizes the tin to a species that will not precipitate a palladium catalyst. Treating the activated circuit with palladium solution results in the selective deposition of palladium metal. The circuit may then be plated in an electroless copper bath.

Another novel technique for creating an additive flex circuit is through catalytic circuit patterns applied directly onto the surface of the flexible substrate by xerographic methods. In this method, a circuit drawing is placed on a specially modified photo copier, charged with proprietary catalytic toners, and copied in the same fashion as normal paper photo copies. The exception is that the paper is replaced by a flexible laminate material. The circuit is then plated by electroless copper deposition. This technique promises to be the most cost effective and offers an extremely short manufacturing cycle. Figure 3.2 shows an example of the process.

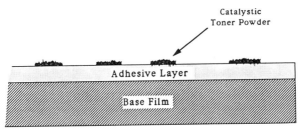

Xerographic Imaging

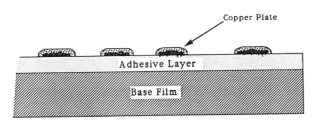

After Electroless &
Electro-Plating

FIGURE 3.2   Xerographic flex circuit

Polymer Thick Film (PTF).   Polymer Thick Film technology is an elegantly simple process for making circuitry. Conductive ink is printed and cured. The PTF method is the lowest cost circuit process in use today because of its simplicity. The technology has specific limits and the designer must become aware of them. This technology is covered in more detail in Chapter 10 on Polymer Thick Film Flex.

PTF conductors are produced by screen printing metal-filled or carbon-filled ink. The silver inks are the most popular because electrical conductivity is best, and the material remains highly conductive in an oxidizing atmosphere. Conductivities of 15 mohms/square/mil (.038 ohm-cm) are common place, and experimental inks have reached 5, but this is still almost an order of magnitude poorer than for values for copper. Yet many circuit and switch designs can accommodate the higher resistance. In fact, higher resistance (10–20 ohm/square/mil) carbon PTF circuits are now routinely used for low end consumer electronics products, such as calculators.

In addition to the limit on conductivity, PTF is generally restricted to conductor widths of 10 mil (25.4 mm) or greater. Although the screen printing process is capable of producing widths under 5 mil (12.7 mm), unacceptable yields result for the printing rates now used by that industry. The process is also limited in layer count. Manufacturers are routinely producing 2-layer circuits, but 4-layer products are done on a very limited basis. Incidentally, two methods are available for adding conductor layers. The double-sided, printed through hole process provides a product structurally similar to the standard copper plated through hole circuit. Ink is pulled through the holes to produce "PTF barrels." The second method borrows from the hybrid circuit industry to produce constructions of "printed up" layers. Dielectric ink is printed over conductor layers with vias or opens, allowing for connections from subsequent conductor layers. The next conductor layer, when printed on the dielectric, prints down into the via to create an interlayer connection.

A major problem associated with PTF circuits has been component attachment. Solderable inks have been developed, but they generally require a rigid, heat resistant base. Because it is desirable to use PTF with low cost substrates, like polyester film, a different assembly strategy was necessary. Solderable PTF inks on thermoplastic polyester are not successful because the ink cannot "stabilize" the base during the heat of soldering as does copper metal. The structure shrinks and curls: a low temperature approach is required.

Developers in the emerging PTF industry looked to conductive adhesives for a solution. Silver-filled epoxy adhesives had been used for decades as die attached materials and good materials were available. Numerous workers in the field reported on all PTF products and processes.[3,4,5] Figure 3.3 shows some of the early designs.

The PTF assembly process involves printing, stenciling, or pneumatically depositing conductive adhesive onto bonding pads, placing components and then curing the adhesive paste. The process and the materials require SMT because strong lap or butt joints can be made and reflow phenomenon, seen in solder but not adhesives, is not required. Feed through device assembly with conductive adhesives is tedious and rarely cost effective because conductive adhesive does not have the necessary high surface tension to wick on to leads and form fillets. The advent of SMT really signaled the emergence of PTF assembly.

3.3 BACK BARED OR DOUBLE ACCESS FLEXIBLE CIRCUITS

Back bared or double access flexible circuits are single conductor layer circuits wherein the metal conductors are accessed from both sides. The technique is applied when component soldering or other interconnection is required on two sides of a low density circuit. This method preempts the need for double-sided or 2-layer circuits and the plated through holes and the extra processing they entail. The thinness of flex base film and the ease of fabricating openings make back bare processing practical. Another advantage of the back bared construc-

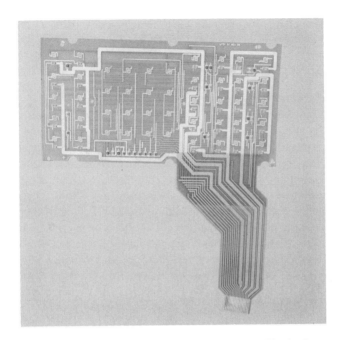

FIGURE 3.3   PTF assembly. (Courtesy of Poly-Flex Circuits, Inc.)

tion is its application to dynamic flex applications. Double-sided circuits cannot be used in continuous flexing situations because the copper will fatigue in a short time. Back bared circuitry is a single-sided construction, and the copper conductors can be placed in a neutral bending axis by using a cover layer film of the same material as the base. This means that the circuit can undergo millions of flex cycles provided rolled and annealed copper is used.

There are several methods for creating a back bared, single conductor layer circuit. They are: pre-punched base film lamination, chemical etching of polyimide, mechanical skiving, laser machining (eximer or CO_2), and plasma etching. Following are more detailed descriptions of these cited processes.

3.3.1 Pre-Punched Base Film Lamination

Back bared circuits produced in this fashion are the most common type. This is because the method offers the best use of normally available technology. The method entails pre-drilling or punching access holes in the flexible laminate film before laminating it to the metal foil that will be etched to create the circuit paths. Following etching, the top cover layer having holes over other termination areas is laminated to the circuit. Holes in the metal foil are generated either by drilling, punching, or during the same etching process that is used to produce the circuitry.

3.3.2 Chemical Etching of Polyimide

This method is specifically for use with circuits fabricated using polyimide films. Polyimide films, while normally very chemically inert substrates, are subject to dissolution in strong caustic or alkaline environments. However, a few of the newer polyimides are resistant to alkali etching. The technique involves the masking of the circuit with an appropriate caustic resistant image either metallic or organic and immersing the circuit into a hot caustic solution. There are a number of different solutions that have been used successfully, but one of the most common is a solution of potassium hydroxide, with or without ethyl alcohol. Because the method normally requires an additional imaging process, it should be selected only when other methods have been ruled out.

3.3.3 Mechanical Skiving

Mechanical skiving is another method for accessing features from both sides of a single metal layer flex circuit. The method is executed by mechanically removing the flexible film from the top (or bottom) of the circuit to create the desired feature. Techniques employed for this purpose have included the following five methods.

Manual Scraping.   While not a very efficient method, manual scraping away of the base film can be effective for a small number of parts. However, it does require a steady hand.

Fiberglass Rod Abrasion.   This method is carried out by chucking an epoxy/fiberglass rod in a drill press and filing the rod to a point of the correct size and subsequently, using this as a spinning abrasion tool for film removal. Success with this technique is sporadic, and the operator's close attention to detail is imperative.

End Mill Machining.   This method is very familiar to machinists. Here a fluted end mill bit of the appropriate size is chucked in onto a drill press and used to carefully cut away the covering film exposing the base metal feature.

Mechanical skiving is not the most cost effective way to double access to a flex circuit design for volume production, but it can be an effective technique for small quantity prototypes or engineering change order.

Laser Machining.   There are two types of lasing that have been successfully used to gain access to metal circuits under cover layer films. The first is traditional hot cutting, using CO_2 or YAG lasers. The other is ultraviolet light photoablation by eximer lasers. Each of these methods has its own advantages and drawbacks. For instance, the former types are very powerful and fast, but they can also easily damage the flex circuit if care is not taken. They also tend to char the material at the edges of the cut. The latter type, (eximer lasers) on the other hand, are less powerful and slower but can also normally produce a much sharper edge and are capable of resolving much finer features. Eximer lasers also tend to require more maintenance and are reportedly more expensive to operate. Regardless of which type is chosen, lasers are superior to mechanical skiving and perhaps on par with pre-drilling techniques for cost effectiveness.

Plasma Etching.   Plasma etching of flex material to achieve back baring is probably the least common method among those that have been discussed. This is due in part to the relative slowness of plasma etching when compared to all the others. On an individual basis this is true, however, when properly prepared for, all access holes can be created simultaneously. This allows plasma etching to achieve some degree of parity with the other methods.

The preparation of a penal for back baring by plasma etching involves the use of a metal mask that protects all surfaces but the areas of interest from attack by the plasma. One method to accomplish this is to coat the surface of the cover film with a thin metal layer and etch openings above the desired interconnect point. Another is to pre-etch a metal mask, and pin it to the flex panel or part before processing. The latter method will not provide quite as crisp an edge definition as the former, but it can be effective none the less.

3.4 JUMPERS/CROSSOVERS

There will be many designs where a single-layer circuit almost does the job, but a few more conductors are needed that cannot be routed on the primary layer. Conductive jumpers or crossovers represent the next level of density increase on our path to multilayered circuitry. This simple concept involves adding secondary conductors to a single-sided circuit. The crossover process is often different than the one that produced the basic circuit. The crossover, as implied in the term, must extend over an existing conductor. This requires that some form of insulation is interposed between the basic circuit and the crossover. There are several approaches to crossovers which will be covered next.

3.4.1 Polymer Thick Film

Polymer Thick Film (PTF) conductors are metal or carbon-filled plastic binders that can be selectively applied in a liquid or paste form and cured to a permanent solid. Crossovers can be produced by printing silver or carbon ink onto and across base circuit conductors. The base conductor that will be crossed over must be first insulated. This is easily accomplished by printing PTF dielectric over the crossing zone. The PTF conductor is printed from one connection point, over the dielectric pad and then to the next connection point. PTF inks can be printed onto copper, solder plate, gold plate, and other PTF conductive inks. The contact area should be clean and tarnish free in the case of copper. Antitarnish is recommended for the base copper circuit.

The design must take into account the higher volume resistivity of silver ink, which is 15–20 times greater. Crossovers are typically used for low current signals. Carbon ink has a resistivity that is thousands of times greater than copper and, therefore, has a very limited use.

3.4.2 Staples

The stapling concept uses a metal wire to make a connection between two conductors. An automatic stapler or stitching machine pierces through insulation and crimps the metal wire to hold it in place. The staple usually enters from the back of the circuit so that the base film becomes the insulator. The staple may also be inserted and wave soldered. Because only one staple is added at a time, this method is only used for a limited number of crossovers.

3.4.3 Surface Mount Technology Components

Surface Mount Technology (SMT) provides one more means of adding density by routing conductors over circuit traces. Small chip style components can be positioned over existing traces. A clever design can use many existing resistors, capacitors, and other components to crossover circuit traces and increase density. Where there are insufficient components in the circuit, low ohm, so-called ''zero'' ohm resistors can be added.

3.5 DOUBLE-SIDED FLEXIBLE CIRCUITS

Double-sided flexible circuits have two conductor layers and, though not as prevalent as single-sided flex circuits, are continuing to gain in popularity among designers and packaging engineers because of the increases in circuit density it affords. Double-sided flex circuits, although more expensive, can be very efficiently produced in dedicated manufacturing lines. Figure 3.4 shows a roll-to-roll or web process line for producing double-sided circuitry. Continuous web lines, with roll-to-roll image placement, plating, etching, testing, and blanking produce millions of double-sided circuits on polyester and polyimide. Double-sided PTF circuits are also in high volume continuous web and sheet production. Hybrids of copper on one side and PTF on the other side, have been used

FIGURE 3.4   Roll-to-roll flex line. (Courtesy of Sheldahl, Inc.)

commercially since the early 1980's. Following are descriptions of the various methods that have been used to fabricate double sided flexible circuits.

3.5.1 Conventional Plated-Through Hole (PTH) Flexible Circuits

The term double-sided strongly implies that there are side-to-side interconnects. Only a limited number of designs, such as high density cabling, do not require electrical connections between the two layers. The most commonly employed method of fabricating double-sided flexible circuits entails the use of double clad flexible laminates and Plated-Through Hole (PTH) technology. The manufacturing sequence is very similar to that used to fabricate rigid double-sided printed wiring with the notable exception that rigid boards do not normally require a flexible cover layer.

A typical manufacturing sequence is to fabricate holes in a metal clad laminate, metallize the dielectric in the holes between metal foils using an electroless copper process, electroplate the laminate to build the plating in the hole to meet requirements, image the laminate on both sides with a positive of the circuit pattern so that the circuit traces and through holes are reliably covered and protected against the etching process that follows, etch the circuit patterns on both sides, strip the resist, and laminate cover layers top and bottom as required.

A variation of the manufacturing theme just cited is to image the panel plated laminate with a negative pattern of the circuit and to plate the exposed circuit pattern with a tin-lead electrodeposit. The plating resist is then stripped, and the circuit etched in an etchant that does not attack tin-lead deposits. The tin-lead may be left in place if a solder-plated construction is desired. This is a very economical process for making double-sided solder-plated circuits. The process is used extensively for producing solder-plated polyester-based flexible circuits where solder reflow methods can not be used.

If a bare copper circuit is required, however, the tin-lead deposit is stripped from the circuit leaving a bare copper circuit identical to the one produced by the first process. The advantage of this process is realized when very little annular ring is present in a design, and the risk of the resist being under cut and etching out the copper from plated-through holes is high.

3.5.2 Double-Sided Circuit Processes with Barrel Resists

Panel plated laminate can be protected from etching by solder-plating as previously described. Dry film photoresist can also be used because this type of resist can be placed over the plated copper barrels: this is known as tenting. However, if the precision and resolution of photoresist is not required, photoprocessing adds more cost than printed ink resist. There are at least three other processes that can be used, although they are not in common use at this time.

The newest one is electrophoretic resist. Electrophoretic resists are newly emerging but based on the older electro-coating process used to apply primers to automotive bodies. The electrophoretic coating is made up of an ionic polymer emulsion that will plate out when current is applied. The panel plated laminate is immersed in the electrophoretic bath and current applied. Both cationic and anionic baths are commercially available. The charged material plates all unprotected metal surfaces including the copper plated-through holes.

Several types of resists are available. Two basic types of photoresist materials are available—negative- and positive-working. The positive-working resist is best suited for double-sided circuits because exposed areas become solubilized and developed away. This means that imaging light does not have to penetrate into holes. Although electrophoretic resists are relatively new, several Japanese, and at least one U.S. company, is marketing the product.

A non-photosensitive electrophoretic resist can also be used. This concept, which may not have been reduced to practice yet, involves screen printing standard resist ink in a reverse image pattern. Next, electrophoretic resist is applied. Because this material will only coat on conductive surfaces, the exposed copper will be coated. The original printed resist is next removed, exposing the copper that is to be etched. After etching, the electrophoretic resist is stripped. The concept requires that one resist be stripped in alkaline and the other in acid or special stripper.

One more concept, similar to the dual resist approach just described, involves the use of a resist treatment solution. Printed resist, in a reverse image, is applied to the panel plated laminate. The panel is then treated with a solution treatment that can react or form a complex with the bare copper. One such compound is imidazole, a material that has been used as an antitarnish treatment for copper. The imidazole should be chemically modified into a derivative that has more chemical resistance to etchants. At least one such material is available from Japan. The imidazole resist can be stripped in acid. This makes it convenient to use standard alkaline-strippable printed resist and alkali etchant.

A final process involves process ingenuity rather than chemical craftsmanship. A standard screen print resist is used in the process, although it may be adjusted for proper viscosity. A panel plated laminate is placed on a specially modified screen printer. The press vacuum base must allow vacuum to be distributed to all of the plated-through holes. This can be accomplished by using a porous metal base or by applying a vacuum defusing material. The simplest method involves applying filter paper to the press bed before placing the laminate. The paper must be porous enough to permit a good vacuum to pull against the substrate.

The resist is printed with the vacuum on. The ink prints the normal pattern image, but where a hole is encountered ink is pulled into the hole to cover the barrel. The ink is dried in the normal manner and resist is then printed on the other side using the desired image. Once again, the ink is pulled into the barrel. Best results are obtained if the vacuum is adjusted to reduce the pull through because a double coating of resist in barrels can be difficult to strip.

3.5.3 Polymer Thick Film Printed Hole

Polymer Thick Film (PTF) circuitry continues to gain in popularity for lower density circuitry. Until the 1980's, virtually all PTF circuits were single-sided, best exemplified by the membrane switch. In 1984, a high volume Screened Through Hole (STH) process was introduced by a leading flexible circuit manufacturer.[6] The patented[7] process involves pulling conductive ink through holes in the substrate dielectric film while simultaneously screen printing the image. Silver or carbon ink can be printed by this method to produce conductive polymer interconnect barrels. A vacuum diffusion web is used to control and diffuse the vacuum. The method has been successfully applied to continuous web, cylinder screen print equipment. Figure 3.5 shows a cross-section of the STH ''silver barrel.''

3.5.4 Copper/PTF Hybrids

The STH process, just described, has been applied to copper circuitry. A diclad copper laminate can be patterned on both sides and holes fabricated. Conductive

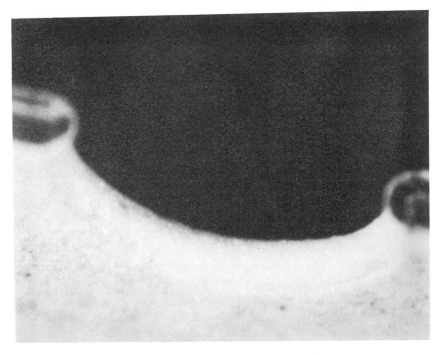

FIGURE 3.5   Printed-through hole cross-section

ink can be printed through the holes to provide interconnections. Alternates include pneumatic and pin array deposition of conductive ink or conductive adhesive. The deposition processes usually produce a filled hole or plug, while the print through method generates a cylinder or barrel.

Another clever variation of the STH process begins with a single-sided copper circuit. Holes are fabricated at the desired locations. The circuit is then printed with conductive ink on the reverse side using the vacuum pull through process. A PTF circuit is produced on the back side which becomes interconnected to the front copper circuit by means of Screened Through Holes of ink. This novel circuit construction offers good cost reduction while offering many of the advantages of a full copper circuit. Components can even be soldered to the copper side. Figure 3.6 shows a copper/PTF silver calculator circuit.

3.5.5 Vacuum Deposited Circuits

Vacuum deposited circuitry has been proposed, and its commercialization attempted for many years. During the mid 1980's, the concept of vacuum depositing copper directly onto flexible dielectric gained popularity, perhaps driven by the need for very fine line flexible circuitry, including TAB. Thin copper

FIGURE 3.6   Cu/hybrid circuit. (Courtesy of Sheldahl, Inc.)

can be etched to finer lines, and vacuum deposition was considered a highly viable starting point for thinner clads. Today, a variety of adhesiveless flexible clads are commercially available. Chapter 2 describes them.

Several developers of adhesiveless flex materials experimented with applying copper and other metals into prefabricated holes. Both sputtering and evaporative deposition methods will deposit sufficient copper into holes for plating. One problem is that most standard hole plating processes will destroy the thin vacuum deposited copper in the holes. Electroplating has been used to plate the typical 1000–3000 Å vacuum-deposited copper. Pattern plating is difficult to control with the thin current carrying layer of vacuum copper. Therefore, panel plating can be used, followed by etching. Two-layer, or adhesiveless TAB, is made by this process.

3.5.6 Non-Plated-Through Hole Connections

Interconnection of the two sides of a double sided flexible circuit can also be accomplished mechanically. Over the years, in fact, a number of methods have been developed. While all of the methods have demonstrated functionality, the individual methods are limited to applications where specifications will permit their use. Following are some of the more interesting and useful of those methods.

Z-Wires. One of the earliest precursors of the plated-through hole, the Z-wire is simply a wire bent in the shape of a ''Z'' and carefully soldered to both sides of a two sided board. This technique is used for vias only as other through hole connections are effected by component leads.

Eyelets. Another ancestor of the plated through hole, eyelets are the equivalent of small rivets with holes at their centers. Like Z-wires, eyelets can make interconnection possible between sides of a two sided circuit. However, unlike Z-wires, eyelets can also accept component leads. In spite of their somewhat archaic nature, eyelets can still be very useful. They can, for instance, serve to interconnect through a stiffener to the flex circuit thus creating a type of low cost rigid flex.

Cold Welding. A newer and potentially very useful technique for interconnecting the two sides of a double sided flexible circuit is by cold welding. The method requires that special tools be fabricated and that circuits be manufactured in a prescribed fashion. Permanent interconnection is obtained by pressing copper into copper through openings in the flexible laminate. Figure 3.7 shows a Z-Wire Interconnect.

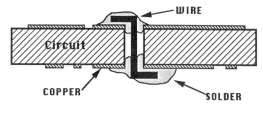

FIGURE 3.7   Z-wire interconnect

3.6 MULTILAYER FLEXIBLE CIRCUITS

Multilayer flexible circuits have three or more layers of conductors. They have, in recent years, become fairly popular as a packaging scheme in spite of their higher cost. Many packaging engineers, in fact, feel they are a bargain for the many benefits they bring, particularly in complicated wiring situations.

Processing of these complex interconnection structures is extremely demanding and requires an exceptional command and control of manufacturing operations. Unlike single and double sided flex circuits, multilayer circuits can only be fabricated in panel form. Along with rigid-flex circuits, which will be discussed later, multilayer flex circuits are among the most expensive types of interconnection structures. Following are descriptions of methods of construction that have been explored to date.

3.6.1 Conventional Multilayer Flexible Circuits

In spite of the title, very little about multilayer flex circuit processing is conventional. Each new design brings with it a host of new challenges. The products are typically engineering intensive and require well thought out and skilled planning. Because each design is unique, it is difficult to describe a typical processing sequence, but the following is offered to provide some sense of the intricacies.

Flexible laminates that are to be the circuit layers are provided with tooling holes. Drilling of tentacle ends may also occur at this time. The layers are next imaged and etched. Cover layers are laminated to the etched patterns providing access to tentacle ends through openings in the cover layer, if required. The cover-coated circuit layers are now laminated together, using flexible bond plies leaving unbonded the tentacles. The outermost copper layers are still coated with unetched copper. The panel is drilled using the same tooling system that was used for image transfer. This is to assure to the degree possible that internal lands will be in register with the drilled holes. The holes are cleaned, using a plasma process or other suitable method. The holes are metallized by electroless copper deposition or other acceptable techniques and electroplated with additional copper to meet requirements. A negative image of the outer circuit pattern is applied to the copper foil. The open areas or circuit pattern are plated with tin-lead. The resist image is stripped, and the pattern etched from the background foil. Next, the outer coverlayer is laminated to the circuits providing openings to the features of interest. A solder coating process may be prescribed at this time, if not the circuit is punched, routed, or cut from the laminate, and the tentacle ends are freed for connector installation if required.

3.6.2 Anisotropic Layer Interconnect

Anisotropic conductive adhesives, also called Z-Axis, are being used to connect flex circuits to rigid PCB's, other flex circuits, and electronic devices. Recently, a concept was developed where double-sided flex circuits were mated together with Z-axis adhesive to construct multilayer circuits. Because the adhesive only conducts vertically, or in the Z-axis direction, electrical connections are produced where opposing conductors from the laminated circuits meet. Design rules require that circuit traces are covered with dielectric where undesired connections would result.

This unusual multilayer construction, called Z-Link™, provides many unique features. The most exciting is the no stress accumulation. In conventional ML circuits, thermomechanical stress is cumulative with layer count since the plated through copper barrel interconnect structure becomes longer with each layer. A point is reached where thermal expansion of the dielectric layers can destroy the copper barrel. In the Z-Link construction, only single-layer plated through holes are required and stress cannot accumulate. This construction method should make very high layer count possible. Figure 3.8 shows a Z-Link construction diagram.

3.6.3 Plated Post Interconnect Multilayers

Plated post interconnect multilayer flex circuits are fabricated by an unusual technique, the method is not too far removed from thick film hybrid circuit manufacturing procedures. In this method, the circuit layers are interconnected by means of solid plated metal posts that pass through from layer to layer. The posts may continuously stack on top of themselves or they may be moved aside to only make connection to one other layer. This technology is best suited to rigid constructions, so it best serves multilayer flex circuit constructions when they can be bonded to a reinforcing base that can be removed later.

A brief example of how such constructions are fabricated is as follows: a specially prepared substrate is coated with a flexible film, where holes are opened to provide interconnect points in the finished product. The surface is next metallized, imaged, and plated up with the appropriate circuit pattern. The surface

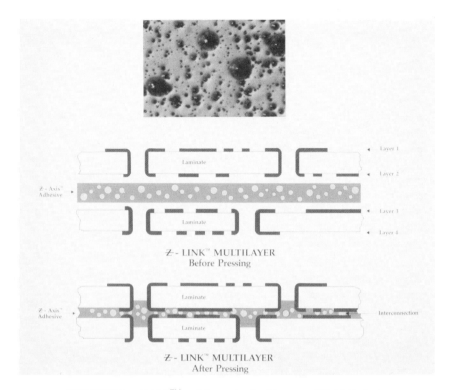

FIGURE 3.8   Z-Link™ multilayer circuit. (Courtesy of Sheldahl, Inc.)

is coated with a flexible film layer. Again, holes are generated in the film, the surface is metallized and imaged, so that the interconnects (posts) only are patterned on the board. The posts are plated up. Next the mask is stripped and the thin surface metallization etched away. The surface may or may not be planarized at this time. Another layer of dielectric film is then laid down, and the cycle starts anew and continues until complete. The circuit is then separated from its base.

This technique allows for very dense structures to be fabricated, but the number of vendors capable of producing this type of product are few.

3.6.4 Printed Up PTF

We have already described the double-sided Polymer Thick Film (PTF) circuit, which has a construction identical to that of the copper PTH double-sided circuit. There is another process for producing double-sided and higher layer count PTF circuits. The process begins by printing a conductive ink pattern on the base dielectric. A dielectric ink is next printed over the conductor pattern in a design that allows interconnects to be made in subsequent processes. Vias, or simple openings, are designed into the dielectric at each point where an interconnect to the first conductor layer will be required. A second printing of dielectric may be applied over the first one to ensure that all printing defects are covered.

A second conductive ink layer can now be applied over the dielectric. Interconnections are created when ink from the second layer flows through the via openings to the first conductor layer. Some processes involve printing a "post" or via fill to provide a more positive connection and to keep a flatter surface profile. This is important for higher layer count. The dielectric printing steps can be repeated followed by another conductor layer. This process is repeated until the desired number of layers is achieved. Workers in PTF technology have claimed up to a dozen conductor layers,[8] but two and four layer circuits are the only commercial products.

3.7 RIGID-FLEX CIRCUITS

Rigid-flex circuits are perhaps the most complex interconnection structures in production today. Having elements of both rigid- and flexible-circuit technology, this type of product brings with it both the best and the worst each technology has to offer. On the plus side, rigid-flex boards provide a superior method for interconnecting the most complex electronic packages. They also offer cost and weight savings over conventional wiring and significantly reduce or even eliminate rework and repair, while enhancing product reliability. On the down side, they represent some of the most demanding technical challenges a flex circuit manufacturer could ever want, and like multilayer flex circuits are extremely engineering intensive.

Though rigid-flex circuits are most often thought of as a military product, more commercial applications are being explored, and many of those are enjoying great success. These successes give positive indications for the future of rigid-flex circuits.

3.7.1 Rigid-Flex Constructions

As with multilayer flexible circuits, there is no typical rigid-flex circuit. Each construction offers its own unique challenges and requirements. In its simplest form, a rigid-flex circuit may be limited to two conductive layers—one rigid, one flexible. In more complex constructions there may be ten, twenty, or more layers of flexible interconnects sandwiched between rigid outer layers. Internal to such constructions might be several of the simpler constructions described above, which serve as interconnect tentacle termination points.

Rigid-flex circuits also serve as a type of formable backplane for system interconnection. These constructions, though normally fairly linear, often require a technique called bookbindering, where each successive layer is lengthened in bend areas in anticipation of the stresses and probable damage that will occur if the technique is omitted. Not surprisingly, the technique is borrowed from the bookbinding industry, which must employ such measures to keep pages flush on a closed book.

Again, as with multilayer flex circuits, there is no typical rigid-flex circuit construction. The manufacturing sequence described for multilayer flex circuits approximates the ones used for rigid flex, with the exception that rigid out layers are used, and more pre-machining of rigid materials is required.

3.7.2 New Materials for Reliability

One significant problem area in the manufacture of military qualified rigid-flex circuits has been meeting the requirements for thermal stress testing and thermal cycling. Thermomechanical stress testing, while demanding enough for standard rigid boards, can be doubly difficult for rigid-flex constructions. This is because a very large percentage of a rigid-flex construction is unsupported adhesives, most commonly acrylic type adhesive.

Unsupported adhesives are known to have large coefficients of thermal expansion, as much as 400 ppm/°C. This translates to excessive strain exerted on the plated-through holes during testing. Thermomechanical failure may appear as a cracking of copper plating at the corners and in the holes. In recognition of this fact, material vendors and circuit manufacturers have worked toward a solution from several fronts.

Low Expansion Polyimides. A very promising approach is to reduce the Temperature Coefficient of Expansion (TCE) for all materials involved. This is not as simple as it may first appear. The flexibility of the dielectric films and the adhesives cannot be reduced. This precludes the use of low expansion fillers used in hardboard materials. Because polymers normally have TCE values an order of magnitude higher than that of copper, unusual polymer chemistry is required to approach this low number. Some of the newer polyimide films, especially the copolymer types, are reaching down in the range of 18 ppm/°C, which is roughly equal to copper.

Polyimide adhesives, with similar low TCE values, are now beginning to replace the high expansion acrylics. These adhesives are more difficult to work with and usually require higher curing temperatures for longer periods of time.

Adhesiveless Laminates. A more direct approach to reducing the Z-axis expansion has been the elimination of the laminating adhesive. A number of adhesiveless copper polyimide products have been introduced and are being slowly moved into multilayer and rigid-flex products. There are a number of manufacturing problems that are created with adhesiveless materials. Many of the manufacturing processes behave differently without adhesives, and considerable development has been required to get adhesiveless materials to work at all. Chemical etch back, for example, does not work well without the adhesive layer.

3.7.3 Designs for Reliability

Another approach to the problem has been through the use of new or modified construction techniques. Among the methods that have been employed with some measure of success are the following:

Reduced adhesive use has been recommended as one approach. This is accomplished by either using a lesser number of adhesive plys in construction or by using bond plys with thinner adhesive coats. It is important to note, however, that this method requires very good control of the lamination process as the filling of gaps will normally be a much more difficult task.

Heavily plated copper through holes have also been recommended as a method for improving reliability of rigid flex circuits. The objective here is to create an extremely robust interconnect that will withstand the rigors of thermal stress and thermal cycle testing. Two things to note here are: (1) that good control of the copper plating operation is imperative. Thick copper that is of poor quality is not going to accomplish the objective. And (2) that the design may require modification in order to accept the additional copper thickness. Oversized holes will be required, mandating the need for larger internal and external lands if minimum annular ring requirements are to be met while accommodating the excess copper.

3.8 INSERT MOLDED 3-D

Flex is the original 3-dimensional circuit. The majority of flex applications take advantage of flexible circuit's ability to be shaped into multiplaner configurations. Some constructions use stiffeners or backer boards to selectively rigidize specific areas. Many designs apply the flex circuit to 3-D molded plastic housings. A good example is the automotive cluster circuit.[9]

In recent years, a number of large and small companies have developed plastic molding processes and introduced plastic molded circuit products. A large number of articles have appeared describing these processes and products over the last five years. One of the more successful approaches involves two shot molding with a combination of platable and inert plastics. The molded part is

FIGURE 3.9   Molded flex. (Courtesy of Sheldahl, Inc.)

plated up with copper, wherever the platable plastic is exposed to create the circuit pattern.

Although several of these processes have been commercialized, there are problems associated with each. A basic flaw is that the molded circuit maker must become an expert molder and circuit processor. While we can all applaud progress, new method strategies should never ignore existing technologies, for eventually present state-of-art always competes with emerging technology. This is the case with molded circuits. Flex was overlooked by all but a few in the early days of molded circuits. Marketing departments briefly mentioned flex as one of the products that would be replaced by molded, not fully understanding the manufacturing advantages and the application features of the flexible circuit.

Long before the molded circuit revolution of the early 1980's, flex users were already using insert molding flexible circuits. Plastic insert molding had been developed several decades ago by the plastic decorating industry. Thin sheets of plastic are printed, finished, shaped, and finally placed in a mold to become an integral part of the molded part when the injected hot plastic mates with the inserted sheet. Flex circuit users simply adopted the insert molding process as a means of rigidizing flex circuits and, thereby produced 3-D molded circuits.

Two basic approaches have been used, trapping circuits in the plastic structure and forming an adhesive bond between the plastic and the flex base film. The trapping design has been used with thermoset circuit materials, especially polyimide, where a plastic fusion bond is difficult. Figure 3-9 shows a part produced by Capsonics for General Motors.

The beauty of insert molded flex is that the plastic molder does not have to learn circuitry processes, and the circuit maker does not have to become a molder. Each expert can concentrate on his own field of expertise. The In-Flex process has not really been exploited, however, perhaps because flex users are content with simply bonding a tested and assembled circuit to a molded and finished housing. There are obvious yield advantages in keeping the technologies separate. Another problem may be the existence of patents covering this insert molding process. Several patents were issued in the U.S. in the late 1980's. This field has been extensively reviewed, and numerous articles have been written.[10, 11, 12] Regardless of any prior art, these patents have not been tested in court for validity and can represent a bar or perhaps an opportunity for a prospective user of In-Flex.

References

1. Gilleo, K. Using Surface Mounted Devices on Flexible Circuitry, *Electronics*, pp. 20–23, March 1986.
2. De Angelo, M. Method of Generating Precious Metal Reducing Patterns, U.S. Patent 3,562,005, Feb. 9, 1971.
3. Gilleo, K. PTF, SMT and Flex, A Winning Combination, presented at NEPCON WEST, Anaheim, CA, Feb. 1985.
4. Gilleo, K. SMT-FLEX—The Synergistic Solution, EXPO SMT '88, Nov. 1988, Las Vegas Hilton, Las Vegas, NV.
5. Gilleo, K. SMT & PTF: Getting It Together—Finally!, EXPO SMT '89, Nov. 1989, Las Vegas, NV.
6. Gilleo, K. Screened Through Hole Technology, *Screen Imaging Technology for Electronics*, pp. 18–22, Feb. 1988.
7. Gilleo, et al. Screened Through Hole Interconnect Process with Plated PTF Ink, U.S. Patent 4,747,211, 1988.
8. Keeler, R. Polymer Thick Film Multilayers: Poised for Takeoff, *Electronic Packaging & Production*, pp. 36–38, 1987.
9. Gilleo, K. 3-D Molded Circuits: Are they Really Here?, *Electronic Packaging & Production*, pp. 112–114, April 1989.
10. Chin, S. Molded Circuit Assemblies at the Starting Line, *Electronic Products*, pp. 33–37, July 1988.
11. Frisch, D. Plastics Add New Dimension to Electronic Components, *Materials Engineering*, pp. 34–37, August 1988.
12. Bahniuk, D. New Way to Build Circuit Boards, *Machine Design*, pp. 99–102, August 21, 1986.

4

Design

Matthew J. Sarri
Sheldahl Inc.

This chapter is about *money*. It's about saving *money*. Lots of *money*.
. . . *It's about designing for manufacturability.*

4.1 INTRODUCTION

The decisions made during design can help ensure success, or they can guarantee failure. Before delving into a host of specific design techniques and considerations, the philosophy of design will be discussed. The goal of this process is to marry the needs of the end user with the capabilities of the flex interconnect manufacturer. A truly successful design, therefore, seldom happens unless the customer and the supplier are intimately involved together at the very start of the process. The end user can become familiar with the basic processing of a flex circuit but may find it impossible to anticipate the myriad of details and intricacies that processing can involve and to a flex manufacturer can become familiar with the basic needs of the customer but may also find it impossible to anticipate the intricate details that the end user faces in finding an optimum packaging solution. The following precautions will help you succeed in finding the perfect design:

If you are an end user, select a manufacturer carefully, and involve him before putting pen to paper. Calling the manufacturer after the print is drawn means you have lost an opportunity to benefit from the manufacturer's design input. If you are a manufacturer, select your customers carefully and work closely with them before deciding. You should define together what it is you are trying to accomplish. What are the physical constraints of your package? What are the electrical requirements? What are the end performance requirements? The end use environment? The reliability issues? The installed cost goals? The tradeoffs? Neither the end user nor the flex manufacturer can adequately answer these questions alone. A close partnership must be developed.

Partnerships are important, but so are the fundamentals of designing flex. The remainder of this chapter is devoted to various technical aspects of design, each of which is important enough to warrant its own separate section.

4.2 THINKING 3-D

A good design integrates electrical and mechanical packaging requirements. Flexible printed circuits can offer multiple advantages to the designer. Most (if not all) electronic packages have spatial requirements in multiple dimensions. Flexible circuitry offers a tremendous design advantage over rigid PCB's because it can be folded and installed in multiple planes, thus utilizing all three dimensions available to the designer without requiring special connectors to change planes. Successful design involves a basic change in spatial thinking from area to volumetric. Integrating multiple planes into the interconnection component, therefore improves the value and reliability of the design. Flexible circuitry also offers advantages over discrete wiring, such as reduced weight, less space, repeatable and reliable interconnection, and in complex wiring schemes reduced cost. Finally, flex circuits offer an advantage over 3-D molded circuits because the tooling cost is considerably lower, turn around is faster, and yield is higher. The following illustrations show how flexible circuitry can simplify and improve a package. Figures 4.1a and 4.1b show how flex can be used to transform an overwhelming point-to-point wiring nightmare into a clean, manageable flex concept. Figures 4.2a and 4.2b show another before and after flex conversion. Note that the design takes full advantage of the 3-D characteristics of flex with the circuit conformed to the shape of the housing. Flex, the original 3-Dimensional circuit, efficiently reduces the *cost of space*.

4.3 GENERAL DESIGN RULES

4.3.1 Mechanical

Part A: Flexing.   Most flex designs fall into two basic categories: flex-to-install and dynamic (or repetitive) flexing applications. The nature of the application places certain constraints on the design. For optimum results in dynamic flex applications, two fundamental design rules should be followed. First, the copper must be located on the neutral axis or center plane of the flex (see figure 4.2a). Such a design places the least amount of stress on the conductor traces; copper elongation during flexing is minimized (see figure 4.2b). If the

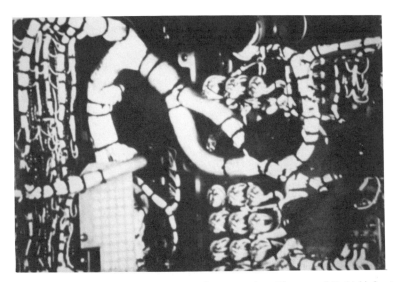

FIGURE 4.1a   Point-to-point wiring before flex conversion. (Courtesy of Sheldahl, Inc.)

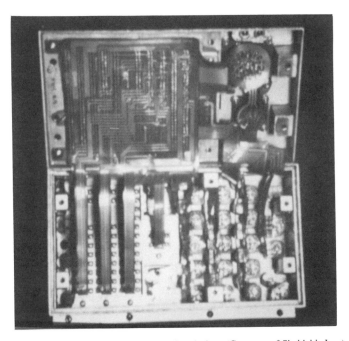

FIGURE 4.1b   After conversion to flex design. (Courtesy of Sheldahl, Inc.)

FIGURE 4.2a   Point-to-point wiring. (Courtesy of Sheldahl, Inc.)

FIGURE 4.2b   Flex conversion using 3-D concept. (Courtesy of Sheldahl, Inc.)

conductors are away from the neutral axis, they will be alternately stretched and compressed (see figure 4.2c) during flexing and will eventually break if the stresses are too great. The second design rule is to use a copper that has sufficient elongation characteristics for the flexing application at hand. This usually means a rolled copper versus an electrodeposited copper unless the radius of the flex is very large. Controversy exists within the flex industry over whether type 5 or type 7 rolled copper foil (ref. IPC-CF-150) should be used for dynamic flex applications. Type 7 or rolled annealed (RA) copper offers the highest elongation value but, it is most difficult to handle in the flex manufacturing process because it is so soft. Type 5 or as-rolled copper is easier to handle, but it has a lower elongation value. There are many factors which can affect the ability of the copper to perform in the final installation including radius of the flex, number of cycles, speed of the flexing motion, degrees of bend, and possible harmonic vibrations or "standing waves" in high frequency flexing. It is difficult to assign hard and fast rules so any dynamic flex design should be carefully tested before entering full production. A low temperature-annealing (LTA) copper foil may also be used to improve laminate handle-ability in the flex fabrication operations, provided the flex manufacturer is able to assure proper annealing of the copper before the flex reaches its final installation.

The copper type and neutral axis design rules place limits on the designer. Ordinary double-sided constructions are often ruled out because conductors will be displaced off the neutral axis. Even if the design eliminates the copper on one side and the cover layer in the dynamic flexing area, the adhesive used for bonding the second copper layer remains. While some designs can tolerate this slight neutral axis imbalance, sensitive or highly critical designs are best left completely on the neutral axis. Special construction techniques are required to eliminate the adhesive from the "back side" of such designs. Even if all these precautions are taken, the copper plating used to interconnect the surfaces via Plated-Through Holes (PTH) also plates up on the surface of the conductors in the flexing areas. Electroplated copper has a much lower modulus of elasticity and, is therefore less reliable in dynamic flex areas. Special dual imaging techniques are required to prevent plating build-up in the dynamic flex areas. Alternate technologies are being developed to overcome these obstacles, but until they are proven reliable, double-sided dynamic flex designs will be much more costly than single-sided flex designs. In some applications, such as high end disk drives, the gains in circuit packaging density and performance justify the extra cost. In most situations however, it is best to strive for a single-sided design.

Flex-to-install applications allow for more options in the design. If the bend radius is sufficiently large, electrodeposited high-ductility (EDHD) copper can be used instead of rolled and annealed. The neutral axis rule can improve flexibility, but in many instances it is not necessary for these applications, and a

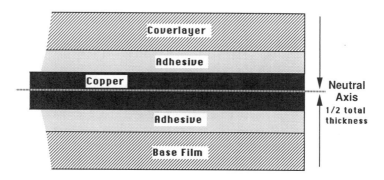

FIGURE 4.3*a*   Neutral axis

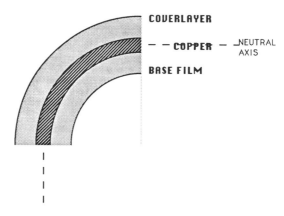

FIGURE 4.3*b*   Neutral axis flexing

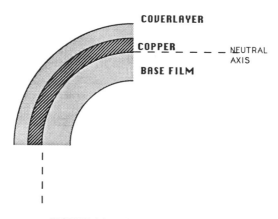

FIGURE 4.3*c*   Non-neutral axis flexing

lower cost liquid cover coat can be used instead of a higher cost cover layer film.

There are various rules of thumb throughout the industry that can be used in making these types of design choices. The best choice for an end user is to work closely with an experienced flex manufacturer to arrive at the design that provides the necessary performance at the best possible cost. *That's designing for value.*

Part B: Dielectric Properties.   Numerous flexible dielectric base films have been historically used to construct flexible circuitry. Chemical resistance, dielectric strength, thermal resistance, dimensional stability, and cost are a few of the factors that effect the choice of a dielectric. Two types of dielectrics are used for the vast majority of flex applications. This discussion will be restricted to the advantages and disadvantages of these two most popular materials: polyimide and polyester film.

Polyimide film is to the flex designer what the Masarati is to the automobile driver. Polyimide offers the optimum in thermal and dimensional stability performance, but it is relatively costly. Polyester, on the other hand, is to the designer what the Escort is to the driver. It cannot withstand high temperatures as well and can be less dimensionally stable, but it is also relatively economical and works adequately in many situations. Polyimide is typically 20 times more expensive per square foot than polyester. The assembly process and the final installation dictate whether this extra cost is justified. For example, some applications may require dimensional tolerances that simply cannot be met using polyester. Other applications may see operating temperatures exceeding the softening point of polyester.

Special techniques for soldering to polyester circuits are also frequently required.[1] The end use and the assembly steps dictate which dielectric is best suited for a particular application. It is wise for the end user to work closely with the flex manufacturer to correctly make this determination.

4.3.2 Electrical

Part A: Current Carrying Capacity.   The amount of current a conductor must carry and the maximum allowable temperature rise determine how large a conductor cross-section must be. The charts in Figures 4.4 and 4.5 show the performance of various widths and thicknesses of copper circuit conductors.

Part B: Impedance.   Impedance can be controlled in the design using one of several basic constructions. The basic constructions and the formulas used to calculate the expected impedance are detailed in articles by W. S. Fujitsubo and others.[2,3,4] The authors have dealt with impedance control and

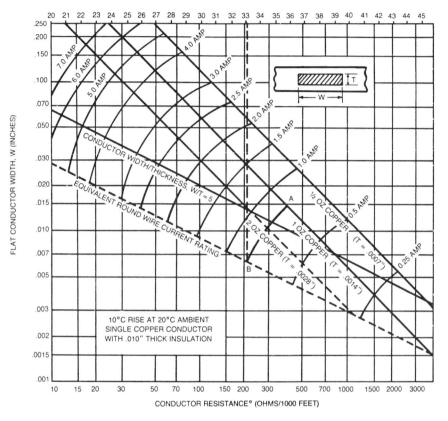

FIGURE 4.4   Current rating nomograph. (Courtesy of Sheldahl, Inc.)

calculation[5,6,7] and software is now available to perform calculations using the basic Greenfield algorithms.

Part C: Shielding.   Ground layers can be used to control cross-talk and RF emissions from a flexible interconnect. The design of the ground layers can greatly effect the noise levels that are emitted. Higher component operating frequencies and stricter FCC emission regulations give this technique significant importance for the future.

Part D: Connectivity.   In order to properly check complex designs, a CAD system should have connectivity and net list compare capability. It should be able to automatically check all the points that have been wired and compare them to a net list to assure that only necessary connections have been made.

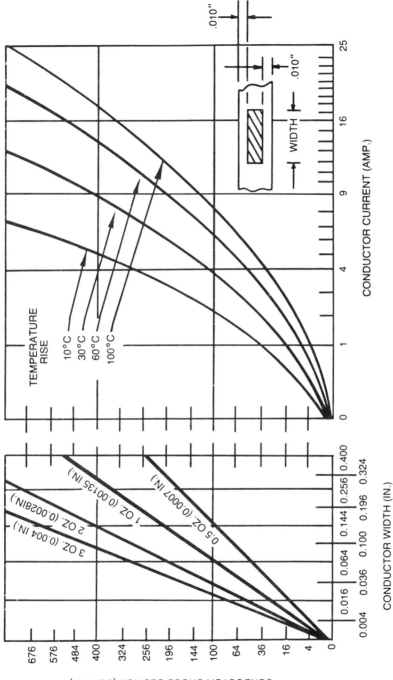

FIGURE 4.5 Current versus temperature curves. (Courtesy of Sheldahl. Inc.)

77

4.4 ROUTING TRADEOFFS

Analyzing the tradeoffs in routing conductors must be a joint effort on the part of the manufacturer and the end user. The end user must define the set electrical characteristics, such as impedance, current carrying ability, resistance, and maximum voltage drop for a given conductor trace. The flex manufacturer can then use this information to arrive at the most cost effective and reliable design. There are many tradeoffs in this step. For example, the designer might choose to design all signal lines at .010″ width and all spaces at .010″. This might provide an adequate trace routing package, but suppose that increasing the trace widths and spaces from .010″ to .015″ would provide a design that could be screen printed instead of photoimaged. This might greatly reduce the cost of the circuit. Let us further suppose that in order to accomplish a screenable design the designer changed from a single-sided design to conductors on the back of the circuit. Would the savings from screen printing instead of photoimaging outweigh the cost of going from single to double sided? How would the option of soldering a single jumper compare with each of these two approaches? Are there components involved in the assembly that could bridge the gap over conductors to increase routing density? Perhaps a Surface Mount resistor could be used as a cross-over. Perhaps low cost PTF "jumpers" could be printed over the copper circuitry to add more interconnects. The answers may vary from manufacturer to manufacturer, depending on their equipment and capabilities.

4.5 ADVANTAGES OF TAPERING AND ROUNDING

Tapering and rounding conductors is a design technique used to enhance manufacturability and to minimize the chance of localized stresses creating conductor cracking problems. Figures 4.6a and 4.6b show the advantage of this technique.

Because conventional hardboards do not bend, CAD software for hardboards typically lacks the aptitude for filleting and radiusing that is necessary to employ this technique. It is important, therefore, to carefully evaluate any circuit board design software if it is to be used for flex. Not all software is ideally suited for this type of design although several packages have been designed specifically for flex.

4.6 CONNECTIVITY: CONNECTORS VERSUS DIRECT CONNECTION

Flex can be used to eliminate connectors completely from a design. See Chapter 8 for integrated features. Also compare Figures 4.7a and 4.7b.

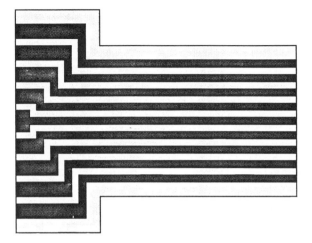

FIGURE 4.6a   Design with no tapering

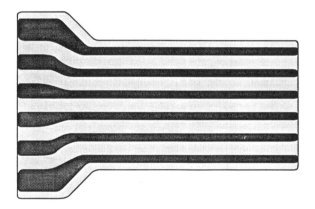

FIGURE 4.6b   Same design with tapering

Figure 4.7a shows a classic rigid board design utilizing four connectors. Components would be assembled to each of the three hardboards and then the connections would be made between the hardboards using ribbon cables or two separate flex circuits depending on the mechanical constraints of the package.

Figure 4.7b shows an integrated package which eliminates the need for the four connectors and combines all the interconnections into a single rigid or rigidized flex component. This illustration is not meant to imply that flex is

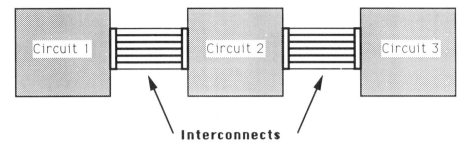

Interconnects

FIGURE 4.7a   Traditional rigid board with cable interconnects

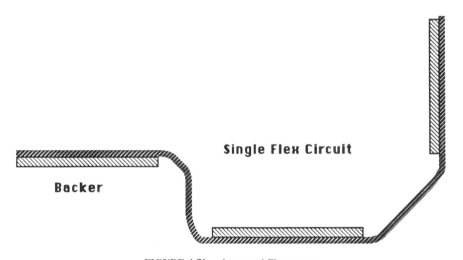

Single Flex Circuit

Backer

FIGURE 4.7b   Integrated Flex system

always the best of the two choices. The complexity of the package, the relative distances of the rigid areas from one another, the vibrational environment, and the densities of the interconnections are only some of the factors that determine the best approach on a particular design. It is important to be aware of the various options available.

Although many connectors can be eliminated through the use of flex, some sort of connection is always needed. A flexible printed circuit is, after all, only part of an integrated interconnect system. There are numerous methods for making the connection from the flexible interconnect to the "outside." Some of the most common techniques and their advantages and disadvantages are listed below:

A. Sculptured Flex
   Advantages:
   1. High reliability: only one connection junction
   2. High vibrational and G-force resistance
   Disadvantages:
   1. High cost of sculptured flex
   2. Difficult to accommodate very fine geometries
B. Lap Bonding
   Advantages:
   1. Low cost of flex
   2. Easy to assemble
   Disadvantages:
   1. Strain reliefs may be required to prevent conductor cracking or pull-away at the interface.
   2. Difficult to accommodate very fine geometries
C. Zero Insertion Force (ZIF) Connectors
   Advantages:
   1. Easy to assemble
   2. Low cost of flex
   3. Can accommodate moderate density interconnections
   Disadvantages:
   1. Cost of connectors
   2. Reliability of connection in severe environments can be a problem depending on surface finishes
D. Crimp-on Connectors
   Advantages:
   1. Easy to assemble
   2. Can be connected and disconnected
   Disadvantages:
   1. Cost of connectors
   2. Resistance to vibration
   3. Pitch is limited
E. Wave Soldered Connectors
   Advantages:
   1. Can be highly reliable, resistant to shocks if proper connector design and assembly is used
   2. Moderately high densities possible
   3. Low cost of flex

Disadvantages:
1. Cost of connectors
2. Cost of wave soldering or hand soldering
F. Surface Mounted Connectors
Advantages:
1. Assembly can be automated
2. High densities possible
3. Low cost of flex
Disadvantages:
1. Cost of connectors
2. Pitch is limited
G. Pressure Contacts
Advantages:
1. Lowest cost
2. Simple assembly and inspection
Disadvantages:
1. Lower reliability
2. Limited current-carrying capability
H. Anisotropic Adhesive
Advantages:
1. Low cost
2. Very fine pitch
Disadvantages:
1. Lower strength
2. Higher resistance
3. Permanent connection
4. Reliability issues

4.7 SPECIAL RULES FOR SOLDERABILITY

It is possible to solder both polyester and polyimide flex. Polyester flex requires a soldering technique that applies heat directly to the copper and creates the solder joint very quickly. If this is not done, the polyester dielectric will melt before the solder joint is formed. It is possible to wave solder polyester flex with special fixturing that isolates the wave to the area where the joint is formed. In some instances it is even possible to solder surface mount components through an infrared reflow operation, but again great care must be taken to shield the non-soldered area of the flex. Polyester soldering requires a highly developed technique and should be seriously considered only if the application volumes are very high[1] (see Figure 4.8).

Polyimide flex, on the other hand, easily survives soldering. The degradation point of polyimide far exceeds normally encountered temperatures. The weak

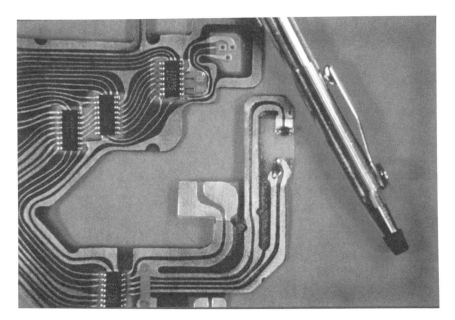

FIGURE 4.8   Polyester circuit wave soldered (Polaroid Camera Circuit Assembly)

link in polyimide flexes is usually the adhesive. With the advent of adhesiveless laminates, this will become a smaller factor in the future. The most important precaution to take when soldering polyimide flex is a pre-bake step to remove moisture from the unit immediately prior to soldering unless a low water absorption polyimide is used. Because many polyimides are highly hygroscopic, they tend to absorb up to 3% of their weight directly from the atmosphere. When the circuit is soldered, the moisture vaporizes within the flex and literally explodes the unit from within. Pre-bake steps become even more important on high count multilayer flex designs. Some of the second generation polyimides absorb less than 1% moisture and do not require a pre-bake.

4.8 LAYOUT FOR MANUFACTURABILITY

4.8.1 The Importance of Reasonable Tolerancing

Why is Tolerancing Different for Flexible Circuits?   A flexible circuit is flexible only because the materials it is constructed out of are flexible. The type of copper utilized is chosen based on the amount of flexibility required for the application. The dielectric films and adhesives are flexible primarily for two reasons: they are thin, and they are unreinforced.

Thinness is a major factor in flexibility because the forces at work during flexing (e.g., compression and expansion) are greatly affected by material thickness. Furthermore, minimum allowable bend radii are a function of material thickness.

The factor that most affects dimensional stability however, is the unreinforced nature of the materials. After all, core materials used in producing hardboard circuits can rival flexible circuit materials in thinness. The difference is that hardboard materials are "supported," or "reinforced" with woven fiberglass or similar material. This makes hardboard materials by nature more dimensionally stable but less flexible than flexible circuit materials. Although very thin reinforced materials can be flexed, flexural fatigue is poor.

What are the Forces Involved?   Dimensional instability is a phenomenon that is observable throughout the fabrication and the life of a flexible circuit. There are two primary kinds of behavior: one is the effect of inducing and relieving stress during the manufacturing process, and the other is the hygroscopic nature of the materials.

When laminates are constructed out of dielectric film, adhesive and copper, stresses are built into the composite structure. The act of laminating disparate materials together using compressive forces and heat cannot help but create inherent stress. During the manufacturing process, selected areas of copper (the most stable part of the composite) are removed as the circuit pattern is created. When this happens, some but not all, of the stresses in the laminate are relieved, often causing shrinkage. Because the adhesive coated dielectric films used in making the laminate are typically produced in roll form, the amounts of stress in the laminate are significantly different in the "machine" and "transverse" directions. Thus the shrinkage or stress relief that occurs at etching is different in the X and Y directions.

After etching of the copper, there are usually one or more additional lamination steps that occur in order to encapsulate the exposed circuitry. By using compressive forces and heat to achieve effective encapsulation, stresses are again built into the structure. This compression causes the circuit to "grow," making it stretch and enlarge. The hygroscopic nature of the materials causes continuous, never ending, reversible changes in dimensional characteristics. These changes can be observed during manufacturing, as each "wet" process subjects the materials to absorption of water and dimensional growth. There are many "wet" operations, such as cleaning, rinsing, etching, stripping, etc. The moisture must be removed prior to any thermal exposures, such as plasma etch, reflow, Hot Air Level (HAL), or solder assembly. Failure to remove the moisture would risk vaporizing the moisture in the circuit, causing it to delaminate or "blow apart," as previously discussed. The effect of removing moisture, of course, is to shrink the material from its water-ladened state.

The process of absorbing (growing) and desorbing (shrinking) is continuous.

Even when a finished assembly is residing in a stock room, aboard an aircraft, or inside a portable computer, the dimensional changes go on. These types of dimensional changes (stress induction/relief and hygroscopic changes) make it easy to see why tolerance requirements must be somewhat liberal compared to rigid products.

What are Reasonable Tolerances?   Tolerances are applied to features. How the features are produced determines what tolerances can be held. In real life, it is often the other way around; the tolerances dictated determine the tooling and manufacturing methods. Unfortunately, this commonly produces two effects: either the circuit cannot be made to meet the specified tolerances, or the fabricator is forced to use expensive and/or low-yielding methods to meet the requirements. Therefore, it is critical that designers tolerance drawings with both the application and fabricator be kept in mind.

Different manufacturing operations will perform within certain dimensional tolerances. For example, a NC drill machine will routinely drill holes within a true position of 0.002". There are other factors that come into play, such as maximum material condition, spindle runout, etc., and so it is safe to assume that a true position of 0.004" can be drilled.

The fact that a true position of 0.004" can be drilled does not imply that it can be held. Recall our discussion of the hygroscopic nature of the materials. If the holes are drilled early in the process, the location of the holes will vary significantly over relatively large distances (6" or more). If the holes are drilled early in the process, the features in the circuit will have moved, and the hole locations will have to be adjusted accordingly. The net effect is an inability to hold normal tolerances over distances. A common rule of thumb is to allow for shrinkage and/or expansion of 0.001" to 0.0015" per inch.

For instance, within a connector pattern a true position of 0.006" is reasonable. You are telling the fabricator that you need to be able to fit your connector into the circuit. Between connector patterns, it is usually wise to expand the tolerance to a true position of 0.010" or 0.014". This should be acceptable for virtually all applications. (Remember, this is a flex circuit.) Over large distances of 6" or more, data dimensioning should reflect the instability of the material. It is no reflection on the fabricator's manufacturing techniques.

The concept is relatively simple when applied in this manner to features created at the same time by the same tool. Another example relates to periphery tolerances. The dimension from one edge to another, when created by a single die, is simply a matter of what tolerance the die can hold. Hard tool dies routinely meet requirements of ±0.005", while steel rule dies are from ±0.010" to ±0.015", depending on the configuration. The only exception is when large distances are involved, as discussed above.

Unfortunately, not all features toleranced on a flexible circuit are created at the same time by the same tool. Hole-to-edge tolerances, for instance, may be difficult to maintain if the holes and edges are created at different operations by different tools. Besides the normal tolerance accumulations due to registration and separate tool tolerances, do not forget that the material has undergone dimensional changes between operations. Rigid-flex circuits, for example, may have flexible edges defined by steel rule dies early in the process, Plated-Through Holes NC drilled midway through the process, and rigid edges routed near the end.

If there are unplated mounting holes in the rigid section, tolerances may dictate when they are created. If they are tied to the Plated-Through Holes, then they should be drilled in with those holes. If they are tied to the circuit edge, they should be secondary drilled during the routing operation. If they are tied to the etched features inside the circuit, they may have to be X-rayed and offset. If the drawing tolerancing scheme does not allow for the possibilities and limitations of the manufacturing process, it may be simply impossible to build the circuit.

Depending on how the tolerances are defined, a manufacturer has certain options available in terms of tooling and manufacturing processes. A designer must be aware that specified tolerances will determine not only the manufacturing methodology, but the costs and lead times as well. The best approach, therefore, is to involve the manufacturer in establishing what characteristics are critical to the flex performance, and how its dimensions should be toleranced.

4.9 NESTING (OR LAYOUT) TECHNIQUES

4.9.1 Shape For Efficiency

Material costs can account for as much as 50% of the cost of a flex circuit. It is, therefore, important to create a design that squeezes as many circuits as possible from each square foot of raw material that enters the process. Most manufacturers have standardized material sizes so it is important to consider size and shape from the outset. Removing just 0.250" from a particular dimension may make a tremendous improvement in the manufacturer's ability to efficiently nest the circuit. Adding a 0.250" can have a devastating effect. Nestability can also improve material utilization considerably. Figure 4.9a and b demonstrate the impact of nestability on material usage.

4.9.2 Directionality Issues

A final consideration in nesting circuits is the direction of the copper grain. Rolled copper has a grain structure parallel to the direction in which the copper was rolled at the copper mill. For dynamic flexing areas, it is critical that the

FIGURE 4.9*a*   Poor nestability

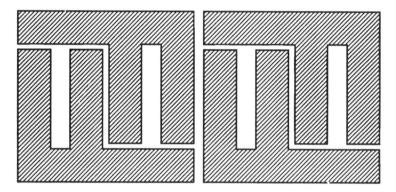

FIGURE 4.9*b*   Good nestability

flexing conductors run parallel to the grain direction of the copper for maximum reliability. This requirement will restrict how some flexes are located in a nesting pattern.

4.10 STIFFENERS

Stiffeners are commonly used to stabilize areas near components and connection points on a flex. They can be made of a dielectric film similar to the dielectric in the flex, or from the same material as rigid printed circuit boards, or they can be made of metal.

4.10.1 Stiffener Fabrication

Stiffeners can be fabricated by drilling, punching, or molding. Drilling offers the lowest tooling cost but the highest production cost per unit. Punching costs more up front for the tooling but provides for a lower cost per unit. Molding costs even more up front, and the fabrication cost can vary with respect to punching, depending on how much money is spent to make an elaborate injection molding tool. The most common reason for molding stiffeners is to provide a very clean burr-free stiffener in critical applications, such as rigid electromagnetic disk drives. Molded stiffeners will be excluded from the remainder of this discussion because they are usually justifiable only in disk drive applications.

As application volumes rise, it behooves the designer to work toward a design in which the stiffener is punchable. Film stiffeners offer the best punchability because they are generally thinnest and do not have any woven reinforcement material in them. The disadvantage of using film stiffeners is they offer the least physical stiffening of the part. Rigid or reinforced circuit board material offers better stiffening action, but it is generally thicker and less punchable. Slivering of this material is also common. These are some of the reasons why film stiffeners cost less than rigid stiffeners.

4.10.2 Stiffener Bonding

The most common adhesive types for bonding stiffeners are acrylic thermoset and pressure sensitive adhesives (PSA's). The latter offer the lowest fabrication cost because the stiffener can be applied through a quick manual or semi-automated tacking step at room temperature. Acrylic thermoset adhesives are more costly to use because heat (above 350°F) and pressure (above 350 psi) must be applied to bond the stiffener. Acrylic stiffener bonding can require special fixturing to keep the stiffeners registered while they are in the press. The need for platen pressing is the main reason why acrylic stiffener bonding is so much more costly than bonding with PSA. Acrylic bonding, however, provides a bond free of any air bubbles which might trap flux; PSA bonds, on the other hand, tend to have tiny air pockets. In the past, this was a major concern for high reliability, high stress applications, but advances in flux formulations and cleaning technologies have greatly reduced this reliability concern. The initial bond strength of acrylic also tends to be higher than the initial bond strength of PSA.

4.10.3 Using Metal Stiffeners For Heat Management

The use of metal stiffeners in the past was often ruled out by the presence of leaded components in the stiffened areas. Component clearance holes in a metal stiffener are difficult and expensive to fabricate, and unless the stiffener and components are perfectly aligned, the component leads can short to the stiffener. The advent of Surface Mount Devices (SMD's) eliminated these problems, and metal stiffeners are becoming much more common in flex designs not only as a means of stiffening but also as a way to provide a heat sinking mechanism for high density heat generating designs. The biggest challenge in this type of design is to achieve a bond between the stiffener and flex which maximizes heat dissipation away from the components. Thermally conductive adhesives have been developed for metal backer bonding. Thermally conductive polyimide is also available. This combination allows metal-backed flex to handle significant power loads.

4.10.4 Using The Enclosure As The Backer

Surface mounting has increased the opportunity for using the enclosure as the backer. The bottom side of the flex circuit can be designed to have no components and also remain free of leads when no feed-through components are used. This permits easy bonding of the flex-SMT assembly to a flat surface or combination of flat surfaces on multiple planes. This can be done by mounting the future enclosure in flat form to the back of the flex, assembling components, and forming the flex and enclosure material simultaneously to create the final enclosure shape. Where this is impractical, PSA can be installed with a release liner to the back of the flex. After component assembly to the flex, the PSA liner can be peeled off, and the flex assembled into the preformed enclosure. This is an excellent method for reducing weight and space needed in a design.

SUMMARY AND CONCLUSIONS

The preceding discussion has only touched the tip of the iceberg of design opportunities with flex. It has demonstrated how many details there are in creating an effective design. The details matter, but the most important messages of this chapter are much more general: if you are an end user, seek out a competent flex manufacturer, and work closely with that manufacturer through design and into production; if you are a flex manufacturer, increase your focus in the area of design. It is the way to provide true value to your customer—the end user. True value is the fundamental basis for a profitable win-win business relationship.

References

1. Gilleo, K. The Mylar Myth, *Screen Imaging Technology for Electronics*, pp. 6–10, August 1988.
2. Fujitsubo, W. S. High Performance Controlled Impedance Interconnect System, Proceedings 7th IEEE/CHMT International Electronic Manufacturing Technology Symposium, pp. 118–120, Sept. 1989.
3. Fujitsubo, W. S. Controlled Impedance Interconnections: Theories, Problems and Applications–Part 2, *Electronics*, Vol. 32, No. 13, pp. 55–56, Dec. 1986.

4. Fujitsubo, W. S. Lithography Impact On Printed Wiring Boards, *Solid State Technology*, Vol. 29, No. 6, pp. 161–164, June 1986.
5. Keeler, R. High Speed Digital Printed Circuit Boards, *Electronic Packaging & Production*, pp. 140–145, Jan. 1986.
6. Harper, C. A., and Stalet, W. *Electronic Packaging & Production*, pp. 56–62, April 1985.
7. Messner, G. Impact of New Devices On PWB Design and Material Selection, *Printed Circuit Design*, pp. 4–13, Nov. 1985.

5

Applications

Keith Casson
Sheldahl, Inc.

5.1 OVERVIEW OF FEATURES

In general product categories (as described in Chapter 3), flexible circuits are designed and used in the following forms:

Single sided
Single sided-back bared
Double sided
Flexible multilayer
Rigid-flex multilayer
Selectively rigidized flex

With these construction options in mind, a thorough familiarity with design alternatives is important in making an optimum wiring choice and in avoiding misapplications. After a wiring method is selected, success or failure often depends on how effectively the designer exploits the advantages of the method and minimizes its limitations. This chapter will describe generic properties and limitations and provide specific case histories.

Some of the advantages of FPC are the following:

- System cost reduction because of eliminated wiring errors, reduction in number of components, especially connectors, and reduced assembly effort.
- Simplified assembly; can be bent around components and bonded to chassis.

- Reduced weight of up to 4 : 1 in weight compared to point-to-point wiring and approximately 10 : 1 compared to standard 0.062″ hardboard.
- Reduced volume of about 7 : 1 and even higher when used in applications that take efficient advantage of multiplanarity and conformability.
- Flexibility; can connect moving machine elements and fold or bend in three dimensions to fit after assembly and/or soldering.
- Improved reliability due to reduced number of connections, reduced handling, reduced mass and stress on connections and joints (this can be a dramatic advantage for surface mount solder joints on FPC).
- Uniform electrical characteristics because of consistent spacing and orientation of conductors and insulation.
- Better heat dissipation due to thin, flat conductors and thin insulation especially when bonded to a heat sink using thermally conductive laminating adhesive.
- Improved packaging efficiencies.
- Compliancy of the thin dielectric is an ideal strain relief for SMT, and thermomechanical degradation of solder joints can be completely eliminated.
- Features can be integrated; i.e. TAB (see Chapter 8).

Some of the limitations of FPC are the following:

- Higher non-recurring cost compared to point-to-point wiring.
- Generally higher unit cost than equal area rigid printed boards (exceptions to this are PTF and high volume polyester based FPC).
- Difficult to change if shape is altered; design must be fixed before tooling to control total costs.
- Limited current carrying capacity compared to round wiring.
- Stiffeners (selective rigidizers) may be required to support large or heavy components.
- Difficult to repair due to difficulty of insulation removal and hand soldering.
- Requires fixturing for automated assembly and mass soldering.
- Impedance can be less consistent than rigid printed boards. However, adhesiveless products can provide superior characteristics.
- Less dimensionally stable than rigid printed boards.

The primary considerations for a good FPC application are *function, cost, reliability, fabrication*, and *improved packaging efficiencies*. A well designed flex circuit in the right application will *reduce the cost of space* and *improve performance*. These considerations may overlap. For example, reliability enhancements and size reduction can also reduce costs.

Functional Considerations. The weight and volume of FPC arrays are typically less than 50% of equivalent round wire interconnections. The flexibility of FPC permits the following:

- Connection of components that have motion relative to each other and allows up to one billion cycles of motion.
- Component insertion, soldering, and testing while the assembly is flat, followed by bending, or folding to conform to three dimensional shapes of case or chassis.
- Return to a flat configuration as necessary to service and repair.
- Opportunity to improve packaging efficiency.

Cost Considerations.   Flexible printed circuits can reduce assembly and installation costs by 20 to 50% compared to round wire interconnections or multiple rigid boards that are connected together by cables:

- Consistent, readily visible location and fixed orientation of FPC conductors and terminals eliminate wiring errors. Simplified inspection, trouble shooting, and rework permit use of relatively unskilled assembly personnel.
- Measuring, cutting, stripping, tinning, routing and lacing are eliminated.

Compared to rigid boards, the lower dimensional stability of typical flexible dielectrics during manufacturing requires additional process and tooling measures to control tolerances. The cost of bare polyimide FPC is generally higher than rigid boards of equal area. However, an FPC array can reduce product costs when used to replace a number of rigid circuit boards and interconnecting cables and connectors. Integrating an assembly with FPC usually reduces other components, terminals, and solder joints saving costs and improving reliability. Polyester-based flexible circuits, especially PTF type, are often lower in cost than equivalent hardboards on an area basis, however.

5.2 FLEX TO INSTALL APPLICATIONS

Flex to install features have been used successfully in virtually all market segments.

5.2.1 Automotive Examples

Instrument Cluster.   Polyester film based copper flexible circuits have been successfully used for volume instrument cluster wiring and interconnections throughout the world for nearly 30 years. These circuits typically distribute power to instrument lighting, various gauges, etc. Most of these circuits, to date, have been single sided. Figure 5.1 shows a typical instrument cluster circuit product, while Figure 5.2 shows the circuit assembled to the case or backcan. Some of the newer designs now utilize two or three single-sided circuits layered upon one another. Figure 5.3 shows a design where two single-sided circuits are mated together. Also, as automotive electrical-electronic instrumentation needs become more complex, with electronic components mounted directly on the flex circuit, double-sided plated-through-hole designs are in development. Figure 5.4 shows a prototype double-sided instrument cluster circuit with plated-through-hole interconnection. Some of these products will be on polyimide film for improved assembly durability and temperature resistance.

The conventional polyester-copper instrument cluster circuits have been manufactured either by mechanically die stamping the circuitry pattern (an embossing process) or by screen printing and etching. Present and future design complexities are reducing the applicability of the die stamping process for these products.

The primary reasons for the popularity of FPC in these applications are installed cost advantages, multiplanar design freedom, and reliability improve-

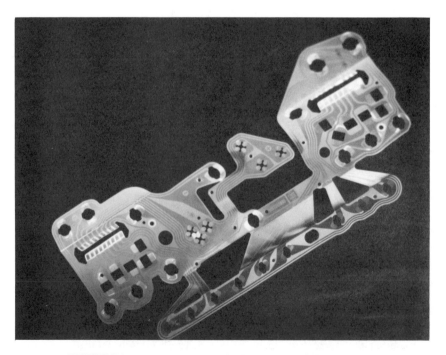

FIGURE 5.1   Instrument cluster circuit. (Courtesy of Sheldahl, Inc.)

FIGURE 5.2   Instrument cluster circuit mounted to housing. (Courtesy of Sheldahl, Inc.)

ments. The cluster circuit is an outstanding, and perhaps the earliest example, of a 3-D circuit produced in volume.

Electronic Controllers.   Complex electronic module assemblies are being evaluated and designed, utilizing polyimide based flexible circuits. Most of these products will be double-sided plated-through-hole designs, utilizing predominantly surface mounted devices. The reasons (major advantages) many of these applications will use FPC are the following:

- Packaging density advantages.
- Excellent high temperature and temperature cycling performance. (Tests have been run showing reliability through more than 2,000 cycles between $-40°C$ and $+150°C$.)
- Thermal control (dissipation) advantages. The thinness of flex constructions, coupled with the possibility of selectively rigidizing the flex with an aluminum or other metal plate for mounting advantages plus improved heat dissipation, gives the most effective thermal control package available.
- Solder joint reliability, especially when FPC is coupled with SMT, can achieve outstanding long term solder joint performance and reliability. Even when the FPC is bonded to a rigidizer, the flexible material layers

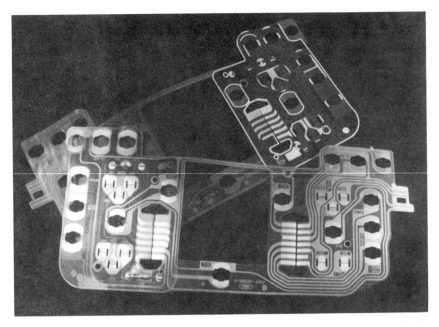

FIGURE 5.3   Two single-side cluster circuits mated together. (Courtesy of Sheldahl, Inc.)

provide a stress/strain relieving function that enables the assembly to successfully withstand thousands of under-the-hood automotive temperature
cycles.

Some of these products are finding use, and others are being considered for
engine controllers, anti-skid braking (ASB), positive traction, transmission, cli-
mate control, entertainment, and numerous other future possible applications.
Figure 5.5 shows an SMT-flex assembly used as a driver circuit and interconnect for an automotive instrument display.

Electronic Sensors.   The electronic automobile of today and future designs
require more sophisticated sensor assemblies. These sensors are used to measure temperature, pressure, rotation, and a multiple of other parameters. In these
applications, FPC offers temperature resistance, fluid resistance, temperature
cycling performance, and conformability. Packaging density coupled with surface mount assembly reliability and environmental (temperature and/or fluid)
resistance, are the keys for these applications. Pressure sensors, designed for
tough, under-the-hood use, are now being produced in high volume. These
sensor designs typically take advantage of the 3-D character of flex and provide

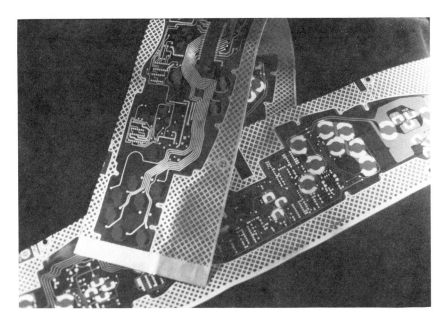

FIGURE 5.4   Double-sided cluster circuit. (Courtesy of Sheldahl, Inc.)

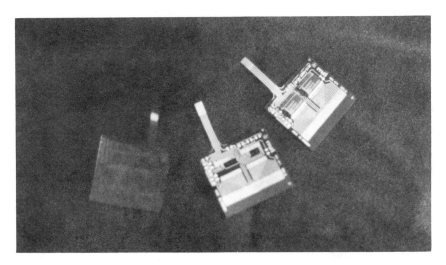

FIGURE 5.5   SMT-flex automotive instrument assembly. (Courtesy of Sheldahl, Inc.)

compactness while increasing reliability. The flex circuit can readily accommodate the popular curved housings used in many sensor packages.

Power Wiring Applications.   FPC is proving to be an excellent and successful option for multiple wiring applications, including valve body assemblies for ABS and positive traction assemblies, multiplex system actuator assemblies, and other systems requirements. Key features can include fluid resistance (i.e. brake fluid), good temperature dissipation, low weight, packaging advantages, and current carrying abilities. Most of these applications utilize single- or double-sided polyimide based copper circuit constructions.

5.2.2 Consumer Product Examples

Cameras.   Most cameras manufactured in the world today utilize FPC. Depending on camera type and complexity, these circuits can typically be either single- or double-sided polyester or polyimide based copper conductor circuits. Some of these FPC camera applications require intermittent flexing requirements (as in folding cameras when they open and close and/or when opening to load film). In these instances, design considerations must include circuitry layout, proper copper and dielectric base, and cover layer/cover coat materials, etc., to ensure adequate life and product reliability. The cameras we all routinely use could not exist in their present compact and functional forms without utilization of FPC. Figure 5.6 shows a polyimide camera circuit in its array form, which can also be utilized for component assembly.

Figure 5.7 shows a folding camera that requires the circuit to flex each time

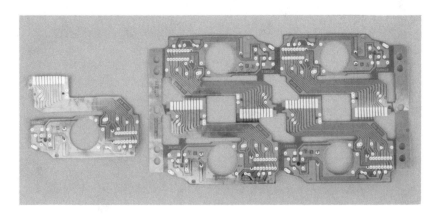

FIGURE 5.6   Polyimide camera circuit array. (Courtesy of Sheldahl, Inc.)

FIGURE 5.7   Camera circuit requiring dynamic flexing. (Courtesy of Sheldahl, Inc.)

the camera is used. The flexible circuit must carry signals as well as the current for the drive motors.

Although many camera circuits are made from polyimide, one U.S. camera maker has used polyester for both feed-through and surface mount devices. Although polyester softens at soldering temperatures, copper-polyester circuits can be soldered. The circuit is designed to maximize the amount of copper so that it serves to restrain the dielectric film from shrinking. Usually the copper is solder-plated so that very rapid soldering is possible. Figure 5.8 shows a camera and its copper-polyester flexible circuit. The circuit is wave soldered. Chapter 9 on Assembly includes a discussion on polyester soldering.

Calculators.   Most hand-held portable compact calculators manufactured throughout the world, now and in past years, utilize FPC. Earlier products, which appeared in the late 1960's and early 1970's, often used more expensive polyimide flexible circuits. Figure 5.9 shows a gold-plated copper-polyimide circuit that was used in a top-of-the line U.S. calculator. Some of these are fairly complex FPC designs. Most of today's calculator products now use low cost Polymer Thick Film (PTF) conductors (silver or carbon screen printable inks) on polyester films. There are still a number of copper-polyester flex circuit designs and combination copper-PTF flexible circuits in use, however. Figure 5.10 shows an all PTF calculator assembly. No copper or solder is used in these

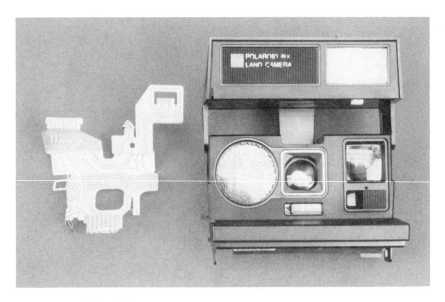

FIGURE 5.8   Polyester camera circuit. (Courtesy of Sheldahl, Inc.)

FIGURE 5.9   Copper/polyimide calculator circuit. (Courtesy of Sheldahl, Inc.)

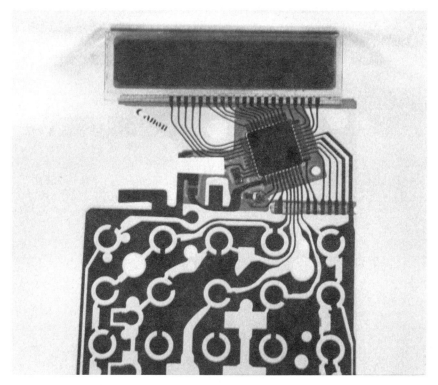

FIGURE 5.10   PTF calculator assembly

products. Chapter 9 on Assembly and Chapter 10 on Polymer Thick Film Flex cover PTF circuit and assembly technology in some detail.

Interconnections are made from the calculator FPC's to batteries, readout displays, logic chips, light powered cells, on/off switches, key pads, etc. using any number of bonding concepts. Solder reflow, pressure contact, anisotropic (Z-axis) bonding agents, conductive adhesives, and remeltable conductors for assembly are used in high volume assembly plants throughout the world. PTF conductors on polyester film are not solderable, but the various polymer bonding methods, just described, work well for these very low cost circuits. In many ways, FPC technology can take credit for being the enabling interconnection technology that has brought us the continuing increase in calculator function coupled with dramatic product price reductions.

Educational Products and Toys.   Most of these products utilize combinations of FPC/membrane switch circuitry for the interface keyboards. These are typically PTF circuitry trace patterns on polyester film materials. Some of these circuits are double-sided PTF constructions with printed-through hole interconnections. The membrane keyboards typically utilize film spacers and fold over circuitry construction. The connections to the logic units are typically made by one or more flexible tails interconnected via connectors or by anisotropic connection technology with heat and pressure bonding. Figure 5.11 shows a PTF double-sided PTF circuit for a popular educational product. A variety of flex membrane switch products are shown in Figure 5.12.

Home Entertainment (VCR's, Video Cameras, Portable Radios and TV's, "Walkman" Radios, etc.). The manufacturing of many of these products is predominantly taking place in Japan, other Asian/Pacific areas, and Europe. Most of these products utilize polyimide-copper, single-or double-sided FPC. Many of these applications also utilize selectively rigidized FPC. One interesting design concept used in a VCR involves selectively rigidizing a copper-polyimide flex with aluminum to create a plug-in module. The aluminum backer has cutouts that allow it to be folded over so that the effective circuit area is reduced by 50%. The use of the metal backer helps dissipate heat and prevent RFI.

FIGURE 5.11   PTF flex circuit for educational product. (Courtesy of Sheldahl, Inc.)

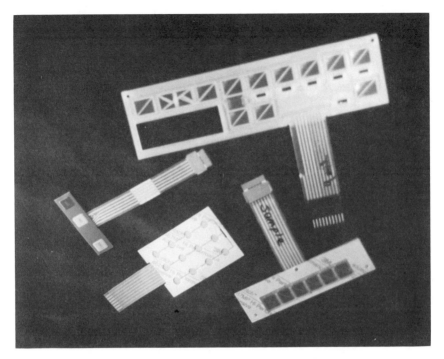

FIGURE 5.12   Flexible switch products. (Courtesy of Sheldahl, Inc.)

5.2.3 Telecommunication Examples

Many telephone handsets have utilized flex-to-install FPC products for packaging advantages, cost control, and reliability for a number of years. These have included a very broad range of products, from the simplest PTF product on polyester film, to more complex double-sided FPC with copper traces on polyimide film or on composite flexible dielectric materials. The conformability of flex makes it ideal for curved handset applications. Figure 5.13 shows how a telephone handset circuit is configured and applied.

More complex business telephones also rely on flex to interconnect the various components. As telephones have become more sophisticated, the benefits of FPC have increased. Figure 5.14 shows an older model rotary telephone designed using a copper-polyimide flexible circuit. The advent of low current display devices, like Light Emitting Diodes (LED's) and Liquid Crystal Displays (LCD's) have permitted the use of PTF circuitry, which has higher electrical resistance and less current capacity than copper. Figure 5.15 shows a PTF-Flex telephone circuit complete with conductive adhesively bonded surface mount components. The circuit is a two-layer design. The circuit density

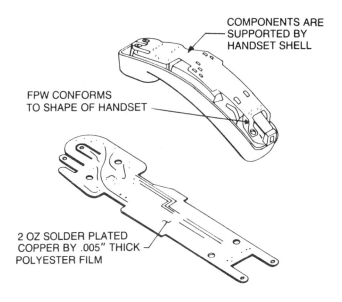

COMPONENTS ARE
SUPPORTED BY
HANDSET SHELL

FPW CONFORMS
TO SHAPE OF HANDSET

2 OZ SOLDER PLATED
COPPER BY .005" THICK
POLYESTER FILM

(A) MOST TELEPHONE HANDSETS WITH INTEGRAL DIAL OR TONE
PADS USE FPW BECAUSE OF RELIABILITY, LOW COST AND
ASSEMBLY SAVINGS.

FIGURE 5.13   Telephone hand set FPC. (Courtesy of Sheldahl, Inc.)

is accomplished by printing dielectric ink over the first conductor layer and then printing the second conductive ink pattern on top. The two layers are interconnected where required by leaving via openings in the dielectric to permit the top ink to form selective junctions with the lower layer.

FPC has also been successfully used in many central office switching applications for telecommunications. A very successful product line has been employed by Western Electric for about 30 years. A special copper/composite laminate was created for this application to handle the demands of greater dimensional stability for flex-to-install situations. Many FAX machines and telecom related products also utilize FPC.

5.2.4 Data Processing Examples

Many data entry keyboards utilize PTF conductors on polyester film to produce very cost-effective membrane switch assemblies. The membrane switch design provides flexibility, cost control, product performance, and reliability. The membrane switch, developed and commercialized in the 1970's, is now the

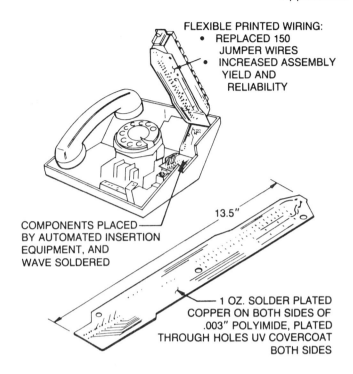

FLEXIBLE PRINTED WIRING:
- REPLACED 150 JUMPER WIRES
- INCREASED ASSEMBLY YIELD AND RELIABILITY

COMPONENTS PLACED BY AUTOMATED INSERTION EQUIPMENT, AND WAVE SOLDERED

13.5"

1 OZ. SOLDER PLATED COPPER ON BOTH SIDES OF .003" POLYIMIDE, PLATED THROUGH HOLES UV COVERCOAT BOTH SIDES

(D) FPW CUT ASSEMBLY COST OF BUSINESS TELEPHONE IN HALF, ELIMINATED RETEST AND REWORK OPERATIONS.

FIGURE 5.14   Older telephone copper-polyimide circuit. (Courtesy of Sheldahl, Inc.)

principle keyboard circuit in use today. Figure 5.16 shows a standard membrane switch product. This product is now evolving into a full electronic assembly complete with all of the circuitry and components to condition and encode the key input to computer-ready information. The encoded keyboard flex circuit is an ideal application for emerging PTF circuit and assembly technology. The use of modern polymer bonding agents allows electronic components to be assembled to the low cost polyester-based switch. Figure 5.17 shows a fully encoded computer keyboard circuit made entirely with PTF materials. The elimination of copper lamination, chemical etching, soldering, and cleaning, saves money and the environment.

In addition, many flex to install FPC polyimide/copper circuits are utilized for interconnection from the logic assemblies to the keyboards. On compact or portable devices, some of these interconnections must withstand intermittent

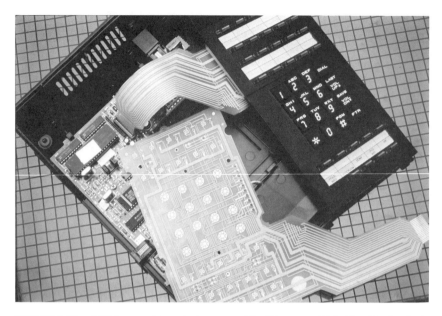

FIGURE 5.15   PTF-flex two-layer telephone assembly. (Courtesy of Poly-Flex Circuits, Inc.)

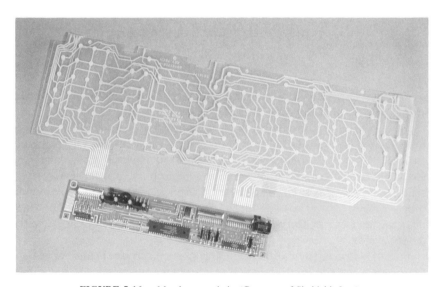

FIGURE 5.16   Membrane switch. (Courtesy of Sheldahl, Inc.)

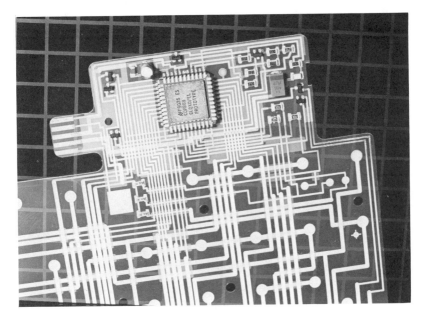

FIGURE 5.17   PTF encoded keyboard circuit. (Courtesy of Poly-Flex Circuits, Inc.)

flexing in use when keyboards are moved relative to the display terminals or on units where the keyboard can be folded up for portability.

5.2.5 Military/Aerospace

A very wide variety of FPC product is utilized for these applications. At the current time, half or more of the dollar value of FPC used in the U.S. annually goes into these market segments. These applications are virtually all on polyimide film with copper conductors and most are flex-to-install applications. Product configurations range from fairly simple high volume single- and double-sided FPC for one time use disposable (i.e., explosive ordnance) applications, up to and including rigid flex designs with as many as 30 circuitry layers for relatively low volume use in military aircraft, missiles, etc. Obviously, these products are quite costly per unit (frequently multithousands of dollars each) and must meet very complex and demanding military specification requirements to attempt to assure the ultimate in product performance and reliability.

The primary reason for FPC use in these applications are weight considerations, packaging advantages (many of these products would not be feasible without flex interconnection technology) and performance reliability. Figure 5.18 diagrams a rigid-flex military application and shows its assembly and use.

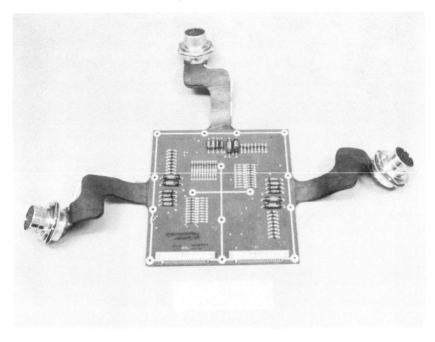

FIGURE 5.18   Military rigid-flex circuit. (Courtesy of Sheldahl, Inc.)

5.3 DYNAMIC FLEX APPLICATIONS

5.3.1 Computer Data Processing

Most true dynamic flex applications are for computer disk drive interconnect systems (to connect the flying read-write heads to the logic assembly), or for printer, typewriter, or copy machine applications to similarly connect the moving head to the logic system. Most of these disk drive FPC designs are of polyimide base film with treated RA copper and polyimide film cover layer. The dynamic flex portion of the flex circuit must be kept single sided. Many typewriter and printer applications have successfully utilized polyester film construction. The disk drive applications have dynamic flex cycling requirements of up to one billion cycles. With the copper traces sandwiched in the neutral axis between matching dielectric layers and the flex parameter maintained at a large radius bend with minimum travel, these applications can function practically indefinitely. Figure 5.19 shows disk drive flexible circuits and assembly hardware.

Printer and typewriter cables are similarly constructed. These tend to be mechanically more demanding flex cycles because of tighter bend radii, longer flex

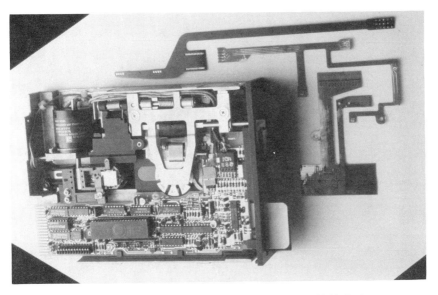

FIGURE 5.19   Disk drive assembly. (Courtesy of Sheldahl, Inc.)

travel, and some mechanical rubbing, but typically 10 to 20 million cycles will meet or exceed the expected life of the product. Virtually all of these dynamic flexing cycles consist of a rolling bend cycle mode with a fairly large bend radius. All of these applications are truly unique engineered system interconnections that could not be successfully accomplished without FPC technology. Figure 5.20 shows an electronic typewriter cable set.

5.3.2 Automotive

An automotive application requiring dynamic flex performance is a "coil spring" type of requirement for air bag actuation systems. These flexible cables tend to be polyester/copper cables with a low number of conductors. The cables may have to have substantial length (up to 80″).

5.4 3-DIMENSIONAL FLEX: REDUCING THE COST OF SPACE

Flexible circuitry is the original and still the most versatile 3-D or "conformal" interconnection system. In the last few years, a lot of money and effort has been expended to develop "molded circuit board" concepts and technologies. There were a number of new venture companies started to develop 3-D molded cir-

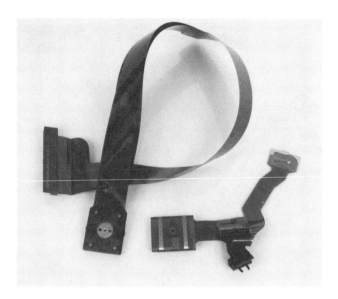

FIGURE 5.20   Typewriter flex cable set. (Courtesy of Sheldahl, Inc.)

cuitry technology. Properly designed and utilized, FPC applications can very successfully and cost effectively fulfill virtually all of the requirements of any of these ''molded board'' concepts. FPC also offers the major advantage of remaining in a flat state during the component mounting, soldering, and assembly stages. After assembly and test, the module can then be bent, formed, folded, etc., to conform to the 3-D shape of the package. FPC should never be overlooked when a 3-D or conformal electrical electronic interconnection application is being considered. Interest in the various 3-D molded concepts appears to be declining with some operations closing or for sale. These technologies had great difficulty trying to replace flex. Consider FPC first, and the applications will usually find successful and cost effective interconnect solutions.

SUMMARY—WHAT TYPES OF INTERCONNECT APPLICATIONS SHOULD CONSIDER UTILIZATION OF FLEXIBLE CIRCUIT TECHNOLOGY

FPC finds its best applications in situations that require or utilize the nature and properties of the flexible dielectric materials, conductive metal foils or pastes, and flexible adhesive systems. These various materials, when properly combined and carefully processed, successfully satisfy many unique interconnect system requirements.

If packaging density, flex to install, dynamic flexibility performance, long term product reliability, weight control, good thermal management and thermal resistance, and/or low stress surface mount solder joints are of value in your application, then a flex solution should be considered. In any application, the more features of flex utilized, the greater the system payoff or return.

Flexible interconnect system technology should not be thought of strictly for cost reasons. Conventional hardboard circuit applications that utilize rectangular circuit boards and have no unusual packaging requirements for bending and no excessive temperature cycling requirements, probably will remain as hardboard circuit products.

Wiring systems that involve widely varying lengths of wires and a wide range of current carrying requirements (or extremely high current carrying capacities) should probably stay with conventional wiring matrices or bundles.

Unless volumes are very high, simple (2 or 3 wire systems) designs of short conductor length should continue with wiring as the set-up, and tooling costs associated with flex will be much higher.

In summary, flexible interconnection system technology should be utilized for those applications where its unique materials and properties help solve a packaging, assembly, reliability, environmental resistance (i.e., temperature, temperature cycling, fluid resistance, etc.), weight, space, or other condition or problem that cannot be easily resolved with more traditional circuitry technology.

6

Implementating Flex

William Jacobi
Jacobi and Associates

6.1 INTRODUCTION

The purpose of this chapter is to alert the reader to the importance of thinking through the intended application of flexible circuitry and working closely with the designer or vendor as early as possible. This process must not be trivialized because it may make the difference between a successful use of flex or an avoidable failure. We will discuss the importance of accurate specifications of the intended use environment, accurate layout of profile, and selection of proper components.

The highest temperature used during assembly and the maximum operating range that the complete assembly will be exposed to, are extremely important parameters for defining the basic type of flexible circuit required. The flex circuit is capable of providing electrical connections throughout a wide range of temperatures. The added requirement of having the flex circuit retain its mechanical properties over this range makes the material selection process of central importance. As with any materials, those used in the flexible circuit are effected by temperature. This means the narrower the defined use temperature the easier job these materials will have providing the desired mechanical properties. Overspecifying the temperature range can add unnecessary cost. The cost and temperature tradeoffs have already been covered in Chapter 2 on Materials and in Chapter 4 on Design.

The flexible circuit should be used to the fullest extent possible to eliminate levels of interconnect. Every eliminated interconnect translates to lower cost and higher reliability. Connectors are a source of signal degradation and potential failure. They need only be used in flex designs where there is a need to have remakable connections or the flex assembly must be connected to the system. Flex is uniquely an interconnect as well as an electronic circuit and assembly medium.

The designer must be careful to consider the final configuration of the circuit in use. Because the fabrication of flexible circuits involve processes that are two dimensional in nature (flat) and usually involve a panel or rectangle in this processing, the designer must lay-out the flex to make sure the processing costs will be manageable. Most flexible circuit manufacturers have capable design services. The trend has been to offer full-blown design services and several manufacturers have set up regional design centers. The potential user who is only familiar with hardboard should take advantage of the expertise of an experienced flexible circuit designer. There are important and fundamental differences between rigid and flexible circuits.

Finally, the selection of components can be important both from the standpoint of mass and termination. As more Surface Mount Devices become available, new techniques are emerging to attach them to the flexible circuit. Chapter 9 on Assembly, describes those methods useful for flex. The designer should be aware of these trends to take advantage of the potential for reducing weight and/or space by incorporating SMT into their designs. The synergistic combination of size and weight reduction provided by both SMT and flex, offers a unique system for miniaturization. Add to this the ability to bend and configure flex into 3-D shapes, and the result is the ultimate in efficient use of space. Flex is clearly the most versatile interconnect and assembly system within use today.

6.2 SPECIFYING THE NEEDS REALISTICALLY

Historically, the most hostile environment that a flexible circuit faced was the various temperatures and chemicals used in processing. This is changing rapidly as new applications are found and less malevolent manufacturing processes are introduced. These new applications are testing the durability of flexible substrates.

Environmental Stress Screening and Use Environments have been developed for the industry and presented to the IPC and its members by Werner Engelmaier.[1] This work describes nine environments that electronic assemblies are subjected to. Because it is important to select and design a system compatible with its intended use, this work is very pertinent. These nine categories are listed in Table 6.1. With each one are the worst case and accelerated test conditions.

The purpose of these tests is to give some predictions of the reliability of solder joints of both through-hole and surface-mounted components. The work can also serve as a guide in material selection and can provide some insight into acceleration techniques.

TABLE 6.1 Nine Categories That Electronic Assemblies are Subjected To

Use Category	Worst Case Environment					Service Life	Service Accelerated			
	T_{min}	T_{max}	δ	t	Hz/yr		T_{min}	T_{max}	δ	t
1. Consumer	0	60	35	12	365	1–3	25	100	75	15
2. Computers	15	60	20	2	14,660	5	25	100	75	15
3. Telecomm	−40	85	35	12	365	7–20	25	100	75	15
4. Coml Air	−55	95	20	2	3000	10	0	100	75	15
5. Industrial	−55	95	20	12	185	10	0	100	100 (& cold)	15
			&40	12	100					
			&60	12	60					
			&80	12	20					
6. Mil Grd & Naval	−55	95	40	12	100	5	0	100	100 (& cold)	15
			&60							
7. Space Geo Leo	−40	85	35	12	365	5–10	0	100	100 (& cold)	15
				1	8760					
8. Mil Avncs	−55	95	40	2	500	5	0	100	100 (& cold)	15
			60	2	500					
			80	2	500					
			&20	1	1000					
9. Auto (Underhood)	−55	125	60	1	1000	5	0	100	100	15
			&100	1	300					
			&140	2	40					

An excellent example of how flexible circuits are being applied in automotive electronics is a polyester based circuit used under the dash. This is shown in Figure 6.1. This industry has used flex as a wiring replacement for decades in the instrument cluster. This is a relatively benign environment (see Table 6.1) from a temperature extreme standpoint. It reduces the thermal exposure even more by using half-turn connectors that make connection by cutting into the exposed circuit pads. This eliminates the use of soldering for the terminations. Thus, the flex is not exposed to any temperatures higher than those needed to cure the adhesives and to dry processing fluids off the surface. This low temperature environment allowed polyester to be used as the substrate.

Certainly, the use of polyester was an economical choice for this environment. The designers peg 185°F as the hottest the interior of a car will get. Now as automotive applications go in front of the firewall where the temperatures can reach 150°C for extended periods, flex must be based on more thermally stable polymers. This has caused the producers of flexible circuitry to consider their raw materials from a new standpoint: service temperature. This property, though certainly not new to most designers of plastic goods, is new to fabricators of flex. The processing environment from the standpoint of heat has always been short term. This has allowed the marriage of one of the many thermally stable polymers: polyimide with a low thermal index adhesive.

6.3 INTERCONNECT ISSUES

A flexible circuit is an interconnect, growing out of the world of wiring harnesses. To effectively accomplish the needs of designers, flexible circuits began to incorporate the new film dielectrics. The flex fabricators learned to use the

FIGURE 6.1   Automotive electronic circuit.

processes of the growing printed circuit industry to build these wiring harness replacements. Using the layout techniques of printed circuit boards, several improvements resulted:

Reduced packaging assembly error
Reduced probability of wiring error
More predictable signal position

Thinner dielectrics allowed several improvements that made the interconnecting task easier:

Increased density through 3-D conformability
Reduced number of solder joints
Weight savings
Reduced space requirements

Finally, when the wiring harness is designed properly it is possible to provide for testing "electricals":

Simplified electrical check and error correction

As the designer scurries around gathering the benefits just defined, some precautions must be voiced. The processing of flex is "aerial" in nature, and the raw materials are expensive. Therefore, efficient utilization of the material and the fabricators process is key to effective flex application. As stated in the beginning of this chapter, the next thing to consider is the layout of panelization of the flex.

6.4 INTERCONNECT OPTIONS

The exact function and/or boundary of the flex interconnect must be defined carefully. The options are almost limitless so the designer must answer the following several key questions:

Is the flex a sub-assembly?
Does it bridge between two or more PCB's?
Are there assembly steps?
Are the mating components defined?
Are there components on the flex?
How will the terminations be made?

Case No. 1   A flexible circuit is hard wired to another unit

In many cases the flex is added to an existing PCB, electrical device, or I/O console. In these instances, the design may not require a connection that can be broken and remade. The flex can then be attached to the device by several means; onto a set of pins or into a set of holes. The attachment of a flex will require an electrically conductive material, such as solder or conductive adhesive. The method used will have a unique thermal process to make the conductive bond/link between the flex and the electrical unit, PCB, etc. The pads on the flex may be exposed to a high temperature for a short time, e.g., 220–260°C for 3–10 seconds to accomplish reliable soldering. When a thermosetting adhesive is used the flex will be exposed to the curing conditions at elevated temperatures for longer times, e.g. 130–175°C for 5–60 minutes. The implications of thermal stress are critical and will be discussed in detail later. The importance of the processing conditions should not be an impediment to using flex successfully. In most instances, this case will have the lowest cost because it will have the fewest parts. Figure 6.1 illustrates the many types of flex that can be used in this case.

Case No. 2   The flex is a sub-assembly, and the termination can be made and unmade

The advantages of a plug-in assembly have long been known, probably dating back to the first wire harness. The flex is repeatedly compared and/or referred to as an improved wire harness so it should not surprise the reader that this is the most prevalent use of flex. The myriad of connectors developed over the years cannot all be used with flex. This means the designer must remember this limitation of flex as the benefits are being exploited. The first question is the issue of assembly sequence and part reduction.

Most connector designs are based on the requirement of mating wires to PCB's. Several have been altered to accommodate flexible circuits. The flexible circuit has fostered several unique styles of connectors because of their thinness or low stiffness. The connectors are usually one or two parts.

One part connectors use the edge of the flex as the male part and can be inserted into a receptacle on the PCB or electrical device. These can be of three styles: a stiffened style, Zero Insertion Force (ZIF) style, and the pressure contact.

A stiffener is usually added to rigidize the flex so the etched conductors can act as fingers. The female mating unit is placed on the PCB via through-hole or SMT methods.

The ZIF style is based on female connector housing that clamps and retains the flex after the electrical contact is made. The majority of the designs in use today are based on mechanisms within the connector structure. One design relies on moving the connector's fingers by a "memory" metal to make the connection.

The pressure connector systems may be confused with ZIF but to the writer's view, most rely on the flexibility or conformability of the flex to make the contact.

Two part multiple-contact connectors are self-contained on each mating unit and consist of paired plug and receptacle assemblies. This usually results in a commonly used grid pattern.

Two part discrete-contact connectors consist of individual male contacts that are attached to the etched conductors on the flex by welding or soldering. These pins can now be inserted into the PCB or next level of assembly.

An alternative to staking, is to form the pins as part of the circuit making process by chemical milling; i.e., selectively etching thicker foil leaving heavier, more robust fingers on the ends of the circuit traces. These fingers can be inserted as pins into a receptacle.

6.5 ASSEMBLY ISSUES AND OPTIONS

The assembly operations that must accompany any interconnection are not limited where flex is concerned. The designer must be aware that the materials take on certain characteristics because of their thinness and low mass. Some absorb moisture and have to be treated in special ways during thermal processing. But as in the case discussed earlier in this chapter, careful planning and design pay off.

6.6 CHOOSING THE RIGHT COMPONENTS

The issue of choosing the right components is raised here not as an admonishment or caveat but as a means of highlighting proper selection. The designer should not feel restricted in any way about the choice of components. The designer is directed to the use of SMC's because of their low mass and, therefore, reduced mechanical stress on the flex. The designer used PCB as shelves for components, flex was used as a interconnect between the shelves. The low mass components allowed area on the flex to be used for components. This is not to say the real estate on rigid PCB is not still attractive. It is just telling the designer that there is another option available. Like everything there are rules, some obvious, some trivial, and some extremely important.

6.7 CIRCUIT LAYOUT

The issue of circuit layout is rarely raised in Rigid Printed Wiring. Layout of the circuit is critical to the effective use of flexible circuitry. The world of flex is hit by this issue in two ways, the free form impact and the consistency whammy. First, the free form impact comes from the thrust by designers to connect everything with one flex. The effect on panelization and material utilization can be disastrous. Flexible material is usually more expensive than FR4 and must be processed through equipment with finite limits. This means an extra leg here or arm there to pick up one more connection may be very costly.

The consistency whammy comes into play in designs requiring high flex life or requiring fine lines. Depending on its intended use, each circuit may have to be positioned on the panel in exactly the same orientation. For example, in high flex life circuits, each circuit should be positioned so that the flexing zone is parallel to the machine direction of the copper foil.

In the case of fine line designs, it may be necessary to allow for shrinkage during processing and, thus, it may be important to place a circuit feature in the same position on the panel each time.

6.8 MOCKUPS

Preparing a model and using a paper doll of the intended flexible circuit is a key step in the manufacturing process. The layout and design of any interconnect must consider the assembly process. Another key function is to extend its connectivity as far as possible. These two items are best done by a model. After making the model, the designer can see those places where relief and reinforcements should be made, and where other assembly steps must be altered. These steps can be best shown in a case study.

A U.S. robotic's manufacturer needed to "wire" the rotating arm in their machine. The steps they took to implement a flexible circuit bears repeating here. The steps are those that any designer should take to ensure the proper use. They had a working machine hard wired, which functioned mechanically in the manner they wanted. It did not function electrically for any length of time. Thus, given the task of improving reliability, the designer made a paper doll cut-out of a flex circuit. Making the cut-out in a shape that unrolled as the arm swung through its arc, the designer could see the interference points and the way the flex moved during the arm's travel. Tying in the appropriate connection points, in this case traditional connectors, he was able to use the paper doll to describe the shape of the circuit. Identifying the area that flexed and the regions that were highly stressed, he was able to position the conductors. Defining the electrical needs of the conductors forced the copper thickness to be two ounces. Though not best from a flex life standpoint, the radius of curvature was ample enough to keep the flex life high.

To reiterate the steps:

1. Make a model
2. Make a paper doll connecting the desired units
3. Overlay the net list
4. Determine the current carrying requirements

5. Position the circuit traces and keep layer count to a minimum
6. Check design rules
7. Identify flex areas, noting which are flex-to-install
8. Panelize—try to reduce material required for each circuit
9. Make a prototype to prove concept

By following this methodology and working with reputable flex circuit vendors, it is easy to correctly use flex interconnections.

6.9 CAD SIMULATION

The sheet metal industry has had an impact on the flexible circuit industry in a very unique way. Just as in the case of bending sheet metal when a flexible circuit is bent, it will take a certain amount of material to do it. Designers of sheet metal enclosures have developed programs to calculate the unfolded dimensions of cabinets. This can aid the designer during the process of making a paper doll. Indeed, it may be possible to lay out the flex completely and accurately on the computer and ensure that design rules are maintained.

6.10 PROTOTYPING

The development of any new product can contain concepts that must be prototyped. Though not an issue in many cases where the item can be handmade and tested, the flexible circuit does not lend itself well to hand-making techniques. The areas of flexibility and electrical performance are just two that can be approximated in handmade models. The flex life of a flexible interconnect is affected by all aspects of the printed circuit process. The uniformity of the etched conductors will have a big impact on how the failure occurs. The conditions of the circuit edge will also be a factor in the ultimate flex life of a circuit.

The electrical characteristics can also be adversely effected by variations in conductor width and dielectric thickness. These two attributes can be effected most by prototype processing.

There are other areas that should be looked at in determining how much money should be invested to achieve a working prototype. We are not saying that prototype flex is not feasible, we are just asking that each case be evaluated on its own merits.

6.11 MANUFACTURABILITY ENHANCEMENT

Flexible circuitry is much more sensitive to manufacturing improvements from design influence than hardboard. Rounded and filleted corners, for example, substantially improve mechanical performance while increasing manufacturing yield. The flexible characteristics of flex mean that process and material direction are important. Unlike hardboard, flexible laminates are made in continuous rolls as well as in sheet form. Some flex manufacturers process in continuous roll form. Circuits are often oriented so that optimum etching and plating processing occurs. Because these processes can be moderately direction-dependent, better manufacturing yield and quality are obtained if the manufacturing engineer has some latitude with circuit layout.

RA copper is extremely directional, and maximum flexural fatigue resistance is achieved when circuits are layed out so that the copper grain structure is parallel to the bending direction. The grain structure always runs in the machine direction of the laminate. This means that high performance dynamic flex circuits, such as those used for disk drives, must be aligned with the copper laminate machine direction.

Some circuits are over coated with screen printed solder mask or protective dielectric cover coat. A better result can be produced if the circuits are angled or skewed in their web layout to permit the printing to occur at an offset angle. Poorer printing quality results when the printing squeegee is exactly perpendicular or parallel to conductor traces. Again, some latitude to angle the parts in the array can greatly improve manufacturability.

Another manufacturability principle, and one that is very fundamental, is to leave as much copper as possible in the design. For example, if a circuit is designed with 10 mil wide conductors and 50 mil spaces, the manufacturability would improve if these conductors could be widened to 20 mils. The circuit would be less costly to produce since the resist pattern could be easily handled with screen printing instead of photoprocessing. Yield would be quite high, and perhaps surprisingly, the manufacturer would use less material. This paradoxical result of using less materials to produce wider conductors is a result of the subtractive etching process. By designing wider conductors, less copper has to be etched away, thus requiring less enchant and less metal reclamation. The basic rule is to leave as much copper within the circuit area as possible since you already paid for it.

6.12 The FINAL DESIGN

6.12.1 Artwork

The responsibility of the artwork should be left to the fabricator, who must build the flexible circuit. Virtually all flexible circuitry manufacturers have good design departments. The flex design rules take considerable time to learn, and standard hardboard CAD systems cannot adequately handle the trace width variance or flaring, so useful in many flex designs. The exception is the larger use of flex with a designer trained specifically in flex.

6.12.2 Tooling

Tooling is a major consideration for flex unlike hardboard that is content with standard rectangular circuit shapes. All flex circuits are cut from an array and, therefore, punch press tooling is usually a consideration. The exception is prototyping or very low production runs where the circuits may be cut by hand, NC knife, or laser. But for high volume production, a mechanical piercing and blanking approach is desirable.

The basic tool types are the steel rule die, which is less costly and quick for turn around time, and the hard tool noted for accuracy and longevity. Within these classes are several subcategories, especially for the hard tool. Each flexible circuit manufacturer will have a slightly different philosophy and approach to tooling, so it is enlightening to understand the tooling to be used on any pending job.

References

1. *IPC Technical Review*, Feb. 1991.

7

Manufacturing Flex Circuits

Steve Peterson
Sheldahl, Inc.

7.1 INTRODUCTION

The basic process steps for hardboard circuits are similar to flex circuits. The primary differences are in the material handling aspects as well as the issues related to dimensional stability during processing. The many advantages that flex gives the end user over hardboard, such as flexibility, can cause major manufacturing challenges to the producer. In addition, building flex in a high volume, roll-to-roll (r/r) fashion, presents other unique problems not normally encountered in the manufacturing of hardboards or sheet processed flex.

The intention here is to give a basic overview of flexible circuit manufacturing and highlight some of the differences encountered when compared to hardboard manufacturing. There are many approaches to building flex. Like every industry, technology is changing and opportunities for automating as well as improving capabilities are available to every flex manufacturer. Information presented in this chapter conveys typical technology and methods. There are a variety of other acceptable methods or technologies not presented in this chapter, however, those presented here may be considered mainstream, "tried and true" methods.

7.2 PROCESS STEPS

Let's start by reviewing the process steps to build a single-sided nonplated circuit. A roll of copper clad laminate, slit to the proper width, is the starting point. These steps, outlined in Table 7.1, offer a broad, generic description for a standard subtractive process sequence.

TABLE 7.1   The Subtractive Circuit Process

Step	Process	Purpose	Method
1	Preclean	Clean Surface. Remove Treatments. Prepare for Image.	Chemical Spray, dip and/or mechanical scrubbing.
2	Image Resist	Define Circuit. Pattern with Resist	Screen Print or Photo Image
3	Etch Conductors	Fabricate Circuit Pattern.	Chemical Spray and/or Dip.
4	Strip	Remove Resist	Chemical Spray and/or Dip
5	Covercoat/ Coverlayer	Provide Electrical and Environmental Protection to Conductors. Provide Discrete Pad Openings.	Platen Press Cover Film, Screen Print Mask, Photomask
6	Cut Holes and Outline	Define Circuit Outline and Internal Holes and Remove Web.	Die Stamp, Laser Cut NC Knife Cu, Drill and Route.
7	Inspect/Test	Remove Defective Product from Lot	Electrical Test, Automatic Optical Inspection, Visual.
8	QC Audit	Statistically Verify Lot Conformance	Random Sample. SPC Data.

Step 1.   Preclean the Laminate.   The purpose of this step is three-fold; 1) to remove the anti-tarnish treatment that the foil supplier deposits on the foil. This antioxidant prevents copper oxidation. 2) to micro etch the surface in order to promote resist adhesion in Step 2. 3) to remove all particulate matter that could effect the subsequent imaging operation.

There are many machine configurations to achieve the process sequence just outlined, whether the substrate is passed through spray chambers or dip tanks or a combination of both.

The basic elements of this process are the following:

a. Dip or spray an acid such as HCl to remove chromate treatment often found on copper laminate
b. Rinse to remove HCl acid
c. Microetch, such as sodium persulphate
d. Rinse
e. Apply antitarnish agent such as benzotriazole or imidazole
f. Dry

The desired result is a clean, dry, chromate-free substrate with a good surface topography to promote adhesion of the resist. While the chromate treatment does an excellent job of preventing the copper from oxidizing prior to use, it is not an ideal surface for resist adhesion or solder plating. It is important to add an in-process oxidation inhibiter to keep the copper stain free during processing. This oxidization could functionally and cosmetically effect the end product as well as cause problems in the subsequent process steps.

The following are critical control parameters necessary to maintain a good consistent process: Chemistry concentrations, temperature of the solutions, and speed of the substrate (dwell time). Precise control of the speed of the material into and out of various baths is critical. Good preventative maintenance on the equipment is needed to maintain cleanliness, good spray pressures, and integrity of the substrate. The harsh chemicals used require extra diligence in the maintenance area. Daily cleanup and maintenance checks will pay dividends in quality and reduced repair.

Step 2. Image Placement of Resist. The two most common methods are screen printing and photoimaging. Xerography, magnetography, and other imaging technologies have been evaluated without much acceptance—at least at this level of development.

Screen Printing. Screen printing today is a more economical process than photoimaging, but it is certainly debatable whether screen printing will be the most cost effective imaging process of the future.[1,2] Continuing improvements in equipment and materials have kept screen printing viable for many applications, even as density demand requirements continue to increase.

Many flex suppliers choose to have only one imaging process rather than invest in the capital and technical support needed to carry both. However, the ability to run both processes allows a wide range of flexible circuit products to be produced economically.

In screen printing, we start with a high precision screen made of metal wire or plastic mesh. A typical screen consists of a rectangular frame (also called a chase) with a meticulously woven mesh stretched over the frame and attached to the frame with clamps or with glue. The mesh is held tightly to the frame under a great deal of tension. This tension is critical for the process to work well, and tensioning is determined with a sensitive tensiometer. The tensioning process is so important that some mesh materials are tensioned in incremental amounts over a 36-hour period. The mesh material is typically interwoven polyester or stainless steel. Silk, from which the old name "silk screen printing" is derived, has not been used for some time. The diameter of the thread can be selected as well as the size of the opening between the woven thread. Mesh count is the number of threads or wires per inch (or centimeter). The smaller diameter threads woven into a finer mesh (small openings) are used where finer printing detail is needed whereas the coarser meshes deposit thicker ink but do not give as much detail reproduction. Available screens allow deposits of less than 5 microns (.0002″) and greater than 50 microns (2 mils) to be deposited.

Once the mesh is stretched onto the chase and attached, it is coated with a photo sensitized emulsion. The circuit artwork is placed in intimate contact with the dried emulsion, and the image is exposed onto the emulsion with UV light. The unexposed emulsion remains soluble and is washed away thereby leaving open areas for the image to be printed.

The printing process consists of registering the screen over the substrate. The chase is held a small distance (.050″–.250″) above the substrate. The printing ink is "flooded" over the top side of the screen so that the print area is fully covered with ink. A squeegee is then pushed downward and across the screen thereby pumping the ink through the mesh openings and onto the substrate. As the squeegee travels across the mesh it brings the mesh and the substrate into momentary contact, the tension on the mesh snaps the mesh away from the substrate immediately after the squeegee travels past the contact point. This peeling away process is critical to ensure a clean, crisp image printed onto the substrate. The printed image is then dried or cured. The ink is now ready to act as an etch resist for the next operation.

Screen printing can be very simple and performed in someone's basement or garage, in this form screen printing is more of an art. Screen printing can also be performed in a highly controlled environment using precision automated printing equipment, in this form screen printing is a science. The successful manufacturer of flex circuit will invest and train to drive their process to a science. See Figure 7.1.

There are two basic screen printing press designs, the common flatbed and the newer cylinder type. The traditional flatbed places the substrate on a flat-table or "bed." The frame and screen remain stationary over the substrate, while the squeegee moves across the print area on the screen. The movement of the squeegee, with its downward pressure, causes momentary contact of the screen with the substrate resulting in image transfer.

The cylinder press carries the substrate, which must be flexible, over the vacuum cylinder in a forward motion. The frame and screen move in synch with the substrate, which is being moved by the cylinder. The squeegee remains stationary at the top of the cylinder. The curvature of the substrate as it conforms to the contour of the cylinder results in a sharper, cleaner image than the flatbed. There is also less screen distortion since the off contact distance (gap between screen and substrate) can be less because of this intrinsic break away feature of curved substrate. Figure 7.2 compares the two types of presses.

One of the drawbacks of screen printing is that there are many critical variables to be controlled. The less a manufacturer is willing to control these variables the more the screen printing process will be compromised. The following is a brief description of some of the critical variables to control.

Phototool. The screen can be no better than the phototool that was used to "shoot" it. The most common types of phototools are silver halides on Mylar

FIGURE 7.1   Automatic R/R screen printing machine

plastic film, silver halides on glass, etched glass, or some other form of deposition on glass. Mylar film is very inexpensive and is convenient to handle and store but is susceptible to dimensional change caused by lack of control over temperature and humidity. Glass artwork has good dimensional stability but is more expensive and more difficult to store and handle. Silver halide is more economical than etched glass but is susceptible to scratching.

The screen.   The printed image can be no better than the screen. The emulsion thickness on the screen effects the thickness of the ink deposit and the definition, or clarity, of the image. Precise coating equipment and techniques are available to maintain a precise, consistent coating across the entire screen as well as from one screen to the next. The tension of the mesh must be tightly controlled, too low a tension will result in inadequate "peel" during printing as well as unacceptable image distortion. Too high a tension will dramatically reduce the screen life. During the exposing process, any dirt or dust present on the phototool will be reproduced onto the screen, also the mesh is normally susceptible to changes in humidity. Thus, the screen should be made and stored in a highly controlled environment, see Figure 7.3. Although liquid emulsion, using a precision coater, is the ultimate for quality, good screens can be made

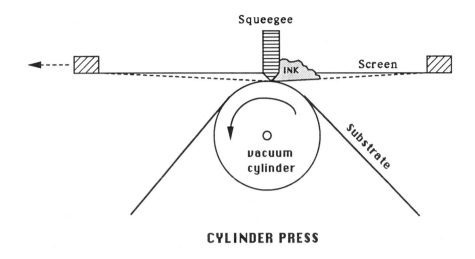

CYLINDER PRESS

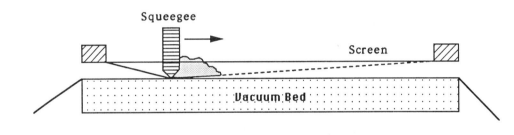

FLATBED PRESS

FIGURE 7.2   Flatbed vs. cylinder press

with dry film emulsions. Film emulsions are available only in specific thickness and, therefore, thickness latitude is restricted.

Machine variables off contact. Off contact is the distance between the screen and the substrate. Too much off contact will result in unacceptable image distortion, too low off contact will result in ink smear due to lack of peel off the squeegee. The durometer, sharpness, print speed, print angle, and downward pressure of the squeegee are all variables that can effect the thickness of ink deposits or the definition of the image. The ink must be controlled, including the viscosity, the age, and the environment. Airborne particles will deposit on the screen or substrate thereby interfering with the printing process. In addition,

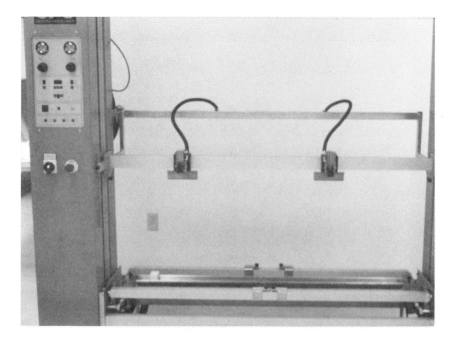

FIGURE 7.3   Automated emulsion coater

changes in temperature and humidity can effect the dimensional stability of the screen and the curing of the ink as well as the viscosity of the ink during printing. Precision screen printing for the electronics industry should be done in a temperature/humidity controlled clean room.

The printing machine. The screen printing machine must be precise and repeatable to allow adequate control and adjustment of the critical variables.

The screen life. A given screen under ideal circumstances will behave in a fairly predictable manner for a certain amount of impressions, but after exceeding its "screen life" dimensional characteristics of the screen may change wildly from the desired nominal. Screen printing involves screen distortion, and the best one can do is to minimize and control that distortion. The manufacturer must determine the maximum amount of impressions a type of screen can safely handle, then have a program in place to monitor screen life and destroy screens when the life is reached.

Ink cure. The curing or drying of ink must be precisely controlled, especially with flex circuits. The ink must be dried or cured enough to have good chemical and abrasion resistance yet not over cured to the point where it becomes brittle and cracks during normal flexing, which occurs during the manufacturing process.

Capability of screen printing. The screen printing capabilities of a flex manufacturer are dependent on its ability to address the variables previously listed. The capabilities have also steadily improved over the years as screen printing equipment, screen making technology, and ink technology have improved. The following capabilities will assume technologies on par with state of the art.

Line width and spacing. 12 mil lines and 12 mil spaces are certainly within screen printings capability with many companies inching toward 8 mil × 8 mil and lower.

Dimensional reproduction. Given that screen printing involves a distortion process, the image on the substrate is never exactly the same as the image of the screen or even phototool, also as the screen wears the image tends to grow in one direction. Assuming a screen with less than 5000 impressions under ideally controlled conditions, a ±.005″ along both axes on a 12″ × 12″ image is achievable. Some of today's screen printing equipment has the ability to optically look at the image and compensate for screen stretch at least in the web direction.

Registration. Often the manufacturer is registering one image to a previously printed image or an image to a prepunched pattern in the web. Again this capability is highly dependent on the type of equipment being used. The most sophisticated screen printing equipment is to use optics to find fiducial targets and to mechanically adjust the chase to good registration. These equipment manufacturers are stating capabilities in the ±.004″ range or less.

Photoimaging.   The object of photoimaging is the same as in screen printing—to place a resist pattern on the copper substrate. The methods to get there are quite different as well as the capabilities.

Photoimaging is a multistep process, with the following steps:

Step 1.   Laminate, coat, or deposit photo sensitive resist.   The most common method today is to laminate a dry film of resist. The technology appears to be heading toward coating or deposition, as the finer the lines and spaces are, the more critical that a uniformly coated thin resist is.

Step 2.   Expose.   The resist is photo activated by UV light through a phototool. The phototool is usually placed in intimate contact with the surface to be exposed. The tool blocks out light where the copper is to be etched away and lets light through where the copper is to remain. The phototool is placed in intimate contact, using a vacuum frame. For finer lines and spaces, collimated light is used. Exposure times usually vary from 3 to 20 seconds.

Step 3.   Develop.   After exposing, desired portions of the resist are polymerized. The portion not exposed, remains soluble and is then washed away in a developing process. After developing, the material has a resist pattern and is ready for etching. A post hardening step, using more UV exposure over the entire area, may be required.

Capability of photoimaging
Capabilities for dry film resist films are stated to be in the .003″ × .003″ to .005″ × .005″ range. Some companies have claimed 0.002″ × 0.002″, using liquid coated resist. The flex manufacturers must make decisions based on the equipment, facilities, and resources that they have.

Alternatives in photoimaging
There are many new technologies that are worth discussing. Among them are:
1. Direct imaging—Computer controlled laser beam exposing patterns directly onto the resist. Among other advantages, no artwork or phototools are needed; the pattern can be obtained by direct interface with a CAD/CAM system.
2. Alternatives to dry film—Resist can be coated onto the substrate by other means, such as electrophoretic deposition, electrostatic coating, and liquid resist coating. A number of processes can be used to coat the liquid resist. One commercial system is described in the next section.
3. Accutrace® System (W.R. Grace)—The system coats resist, exposes, and develops in one line. The resist remains liquid during exposing. The phototool is not placed in intimate contact but within a few thousands of an inch of the substrate. A highly collimated light source is used to maintain integrity of the image. See Figure 7.4 for photoimaging machine.
4. Seriflash® (DuPont)—The phototool is a liquid crystal display.

Step 3.   Etch the Copper.   The resist is impervious to the etchant, which dissolves the copper. The etching process is basically a chemical dissolution method. Copper metal is converted to soluble copper salts. The simplest concept is to dip the resist-coated sheets into an etch bath. Better results are obtained with spray etching or a combination of dip and spray. A well designed spray etch system will provide straighter side walls or trace edges. This is because the spray can produce a more selective etching effect because of its directionality. The spray strikes the copper protruding from under the resist, and thus tends to etch downward instead of in all directions. Dip etching alone, tends to produce a more beveled edge as the etchant "cuts" down and sideways into the copper. The ratio of downward cut (depth) to lateral etch (width) is called the etch factor. The etch factor may vary from 1:3 to about 1:8, depending on etchant and equipment. Spray etching produces the most favorable etch factor. See Figure 7.5 for an etching line.

Step 4.   Strip the Resist.   The resist that defined the image and controlled the etching is now striped away to leave a bare copper pattern of conductors.

Step 5.   Dielectric Placement.   As discussed in other chapters, it is desirable to place a protective dielectric covering selectively over the circuit pattern, thereby exposing only the copper land areas needed for access or assembly. The

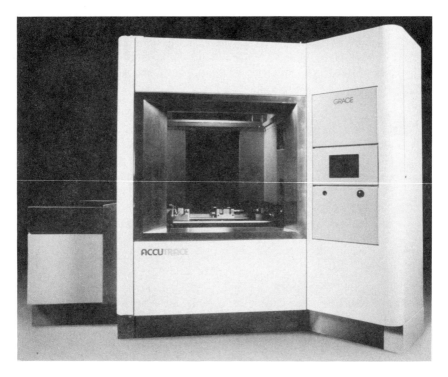

FIGURE 7.4   Accutrace® system. (Courtesy of W. R. Grace Co.)

remainder of the copper traces are insulated from the outside environment and from each other (particularly if the circuit is to be folded).

There are several methods to provide this function. Each method gives the end user advantages and disadvantages, which were discussed in Chapter 2 on Materials.

The first method discussed is the cover layer process, where a prepunched insulating film is bonded to the circuit. This process has the following three steps:

1. Form the Dielectric Layer. This can be done by die cutting, laser cutting, NC drilling/punching or other means of machining, or punching the hole patterns into the dielectric layer. The dielectric layer usually consists of a film coated with an adhesive and covered with a protective release film, that is removed later in the process. Depending on the hole formation process, there can be a "deslugging" step, whereby the scored or punched areas are removed from the substrate.

2. Align Dielectric Overlay with Circuit Pattern. The release liner discussed

FIGURE 7.5   Etching line. (Courtesy of Sheldahl, Inc.)

above is removed. (Some types of adhesive do not require a release film.) The precut film is aligned with the circuit pattern by one of several methods. First, it can be done manually, aligned by eye or with the aid of a mechanical fixture. Secondly, it could be automated. The methods of automation are as varied as ones imagination, and capital resources can provide. Both the cover layer and circuit layer can be in roll form, at this point, and both webs of materials can be registered and laminated roll-to-roll. Robots with vacuum pick-up arms can be used for discrete automatic alignment. If the adhesive system is a thermally cured adhesive requiring relatively long cure cycles the film must be "tacked" or held temporarily in place until the cure cycle. The most common and most flexible method is manually placed overlay and hand taking with a hot iron.

3. Bond the Layers. For thermally cured adhesives, a hot press stage is needed. For high temperature adhesives requiring long cure times, the step involves a hot platen press. For short cure adhesives, the product can be adequately cured in a short nip roll cycle where the assembled layers are run through heated nip rollers. See Figure 7.6 for a platen press machine. The desired result is a product where the dielectric film is adequately conformed to the circuit pattern providing good encapsulation around the traces with adequate bond to the base layer.

Some of the problems that can manifest themselves in this process are the following:

FIGURE 7.6   Platen press. (Courtesy of Sheldahl, Inc.)

- Foreign material—during the assembly or combining process airborne, dirt or particles can be deposited on the circuit surface and, subsequently, be encapsulated into the layers.
- Air bubbles or nonencapsulation—if the heating cycle is not adequate or the conformable materials used to press the layers together are not adequately conformable or several other problems, such as inadequate pressure or poor materials etc., the result can be areas where there is no adhesion (and conformity) of the coverlayer to the circuit layer.
- Adhesive squeeze out—during the heat and pressure cycle, the adhesive flows and can be squeezed out into the open or exposed copper areas, thereby contaminating the surface.
- Misregistration—the coverlayer is not aligned properly to the circuit layer.

The second method of placing a dielectric layer on the circuit is to screen print a liquid in a discrete pattern and harden it. The screen printing process is similar to the primary imaging process previously discussed. The main difference in the screen print method is that the image is produced by the printing step, which is essentially a mechanical image formation process. A liquid or paste pattern is now printed over a surface that is covered with 1- to 3-mil thick traces instead of a flat, smooth surface. The covercoat materials are then cured thermally or with UV radiation. Some of the usual problems encountered are smearing of the liquid during printing, voids in the liquid after printing, and registration of the cover coat image to the primary image.

The third method, uses a photoimagable material as the dielectric protective layer. The process is similar to the photoimaging process previously discussed. The advantage of this process over coverlayer is cost. The advantage of this process over the second process of screen printing the covercoat is the integrity of the image. The photoimagable mask can be imaged and aligned much more precisely than screen printing. The biggest problem with this process is the type of materials normally available do not meet all the needs of the customers. For applications requiring dynamic flexing or folding, today's materials fall short.

Step 6. Die Cutting. The final step in the circuit fabrication process is to cut the outline and internal holes into the circuitry pattern and separate the part from the web. There again are several methods of achieving this result. A hard tool (matched set) die mounted in a punch press can give consistent, accurate patterns in a relatively efficient process. The web of flex patterns can be manually or automatically fed into the punch press machine. The die can be built to a number of configurations, the male and female working portions of the die can be permanently attached to a die set. In this case, setup is very easy as the entire die set is slid into the press and is ready to go. The male/female portions can be mounted onto a die set as part of the production setup process. This results in a longer setup time, but the cost of the tooling is cheaper. For circuits where there are internal holes to generate as well as an outside perimeter cutline, the die can be "staged"; i.e., the die punches the holes in one step then the web is stepped and re-registered to a second stage for the cutline. The die cuts different parts of the same part in different areas of the die. One can have two separate die sets—one to punch the internal holes, and one to cut the outside perimeter. This could then be done at two different stations. One could also build a die to cut the internal holes and the outline in the same step. This type of die is referred to as a compound pierce and blank die. The compound pierce and blank die has the advantage of maintaining a high degree of accuracy and repeatability of cutline to hole dimensions. For manual punching operations, this also tends to be more efficient. The disadvantages of a compound pierce and blank die versus the simple dies, is that since the compound pierce and blank die is more sophisticated it is more expensive and usually requires more maintenance than the other dies. For processes that are automated or semi-automated, the compound pierce and blank die may not be the most economical choice.

Another common die cutting tool for punching out circuits is the steel rule die. A rule die is a series of steel rule blades and punches that are mounted into a base plate that is usually made out of plywood. The punching action is entirely different from the male/female hardtool. Whereas the matched die uses a shearing action, the rule die uses a pinching action as the rule die is pushed against the circuit, which is held on a cutting plate. With the proper equipment and setup techniques, rule die punching can offer clean accurate cutting over several thousand impressions. Depending on the equipment and materials to be cut, rule die cutting can also result in poor, ragged cuts with very limited die life and accuracy. The disadvantage of a rule die over a hardtool matched type die is less accuracy and repeatability. The rule die has a lower quality cut and has far less life than the matched die setup. As stated before, the degree of disadvantage is highly dependent on the type of punch press used, the method of setup, and type of materials used for the tool. The advantage of a rule die is that its cost is usually 1/3 to 1/10 the cost of a hardtool die, and the lead time is considerably shorter. In terms of process efficiencies, the advantage for either type of tool is dependent on the equipment being used. The rule process often only cuts or scores the patterns onto the substrate, and the removal of "slugs" must be performed in a secondary step. The matched dies can be built to provide automatic deslugging.

The third category of hole formation and part cutting can be referred to as tool-less cutting. The cutting is performed by a machine that controls a head which follows a preprogrammed pattern. There are several technologies available: laser cutting, NC knife, and water jet cutting. Internal holes can be formed by NC drilling or NC punching. Rigidized flex can be cut out by NC routing. The advantage of using tool-less technology is no tooling cost or lead time for tooling. Many of the tool-less techniques are capable of more intricate cutting than tools. There are several disadvantages of tool-less cutting: one is that this technology is considerably slower than conventional die cutting. In high volume applications, cost and capacity can be a consideration. The tool-less cutting may not have the cut quality of matched hard tool cutting. Laser cutting tends to leave charred edges, and water jet cutting tends to look slightly ragged.

The ideal application for tool-less cutting is low volume applications with quick turnaround requirements. See Figure 7.7 for a die cut tooling.

Modern die cutting presses now can handle both sheet and web automatically. Several registration methods are in use today ranging from edge and index holes to full vision systems. There is a strong trend toward automatic vision presses, especially for roll-to-roll lines. Cameras read feducials and adjust X, Y, and 0. These presses can operate at very good rates with better accuracy than manual mechanical registration.

Step 7. Verification. The three most common methods to verify circuit functionality are 1) electrical test, 2) automated optical inspection, and 3) human, visual inspection.

Electrical Test.   There are many types of equipment available to electrically test the circuit. There are two parts of the tester to discuss; the system that controls the testing, and the test fixture. The system that controls the testing

FIGURE 7.7   Die cut tooling. (Courtesy of Sheldahl, Inc.)

may have sequential measuring of resistance from circuit point to circuit point. This system will step through a preprogrammed sequence of voltage checks to determine whether the circuit pattern has undesired opens or shorts. The tester is usually capable of applying voltages of varying degrees. Testers also vary in their capability to measure resistances, depending on the threshold capability of the tester. A tester may only be capable of measuring a short with a resistance of 1 megaohm or less or have limited capabilities in setting the threshold of what an open circuit is. The following is a list of features or issues concerning an electrical tester:

- Current setting short circuit
- Voltage setting, open circuit
- Programmability or autolearning capability
- Resistance measuring capability
- Diode testing
- Dwell time
- Capacitance and inductance testing

The other part of the electrical test system is the test fixture. The typical test fixture consists of a test bed with spring loaded pins positioned to align with the desired circuit test points. The spring loaded pins can be of several sizes and configurations, each designed for a different test pad size. A clamping system is needed to place the circuit pattern in intimate contact with the spring loaded probes. Some test systems have 2 NC or robotically controlled test pins, which verify a circuit by stepping through each pair of points and verifying continuity. See Figure 7.8 for automatic r/r electrical tester.

Automatic Optical Inspection (AOI).   Each year, major advances are made in Automatic Optical Inspection (AOI). Such inspection systems are quite commonplace today. The typical AOI system optically checks a circuit pattern per a preprogrammed set of rules and/or patterns to see if the checked pattern conforms. If a discrepancy is found, the AOI system will usually point out the discrepancy to an operator at a verification station. The AOI systems can check for much more than functionality. It can look for conductor width and spacing violations, surface anomalies, and much more than an electrical tester can detect. The one verification that most AOI systems cannot make is through hole continuity. See Figure 7.9 for an AOI system.

Human Inspection.   Finally there is human inspection. This type of inspection requires no capital investment but is the least reliable of the methods used.

In reality, many manufacturers will use all three of the verification methods in their fabrication process, described in Step 7.

Step 8.   QC Audit.   Although today's world class quality design philoso-

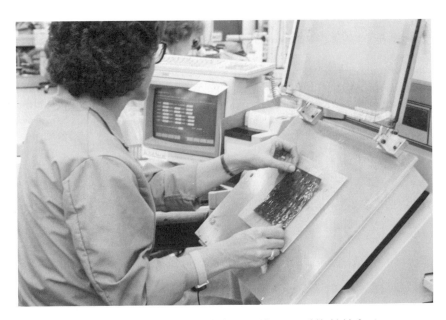

FIGURE 7.8   Automatic electrical tester. (Courtesy of Sheldahl, Inc.)

FIGURE 7.9   AOI system. (Courtesy of Orbot, Inc.)

phy is directed toward designing quality circuits and statistically controlling the process, zero defect processes are rare enough in flexible circuitry manufacturing to require QC audits. Audit information is valuable if not essential, for obtaining the feedback to improve quality and the overall manufacturing process. The traditional QC audit involves sampling the product after various processes, to arrive at yields and sigma limits. The advent of modern automated equipment now makes it possible to do automatic audits. Not only do testers have the ability to record data, but robotic assembly equipment can test components and issue QC reports. A state-of-the-art pick and place machine, for example, can test components before assembly, log results and even switch to a different batch of components, depending on the preprogrammed criteria.

7.3 Other Common Processes

While the above sequences may describe the most common processes in manufacturing a flex circuit, there are many more that are worth mentioning.

Plating.   A circuit pattern of copper traces can be plated over with metals or alloys. This can be done electrolytically or with an electroless process or a combination of both. In the case of a multilayer circuit, the various layers that are laminated together must be electrically connected by plating the barrels of through-holes or vias. Some of the more common metals plated onto conductor traces include, but are not limited to, the following:

- copper
- nickel
- gold
- tin/lead

Solder Reflow.   Circuit traces can have molten solder applied to the surfaces. This can be done in a number of different ways, such as dipping into a molten solder bath and then leveling with hot air or with heated oil. Circuits can also be passed over a wave of molten solder, although this may leave too heavy a deposit. A more precise method, called roll tinning, is to pass the circuit over a transfer roller of molten solder. Solder can also be applied as solder paste, followed by reflowing. This method can be used where paste is desired only in certain areas. Selectively tinning can be accomplished by simply soldercoating pads by hand iron.

Discrete Assembly.   Various stiffeners, brackets, backers, and adhesives can be assembled to a flex. The methods and machines are too numerous to mention in this chapter.

7.4 ADDITIVE CIRCUITRY

The chapter thus far has described the subtractive manufacturing process where a sheet or layer of copper is selectively etched away to form the circuit patterns. An alternative process, which is gaining in its use and may some day totally replace subtractive circuitry, is additive circuitry. In additive circuitry, we start with a dielectric layer and "grow" or plate the desired circuit pattern directly onto the dielectric surface. This process offers a product with several inherent advantages over conventional subtractive circuitry. Additive circuitry processes allow finer lines and spaces and better trace uniformity for impedance control. In addition, the additive circuit has no adhesive layer and can therefore have better heat and chemical resistance as well as offer a circuit with outstanding optical clarity. And finally, since the copper is put down only where it needs to be, there are inherent material cost advantages as well as waste disposal advantages. As attractive as the additive process seems, there are still several disadvantages that prevent many manufacturers from using it. First it is a new and emerging technology that not every circuit manufacturer has knowledge of. Secondly, the technology is often expensive. One serious obstacle, however, is that the resulting copper is electrodeposited (ED) and will not take continuous flexing like rolled copper. Additive circuitry is thus precluded from use in dynamic flexing applications. Progress in electroplating and low temperature annealing may change this situation in the future. Applications requiring high current and thicker copper do not show an inherent cost advantage for additive circuitry.

Stamped Circuitry.   A very simple mechanical circuit patterning process was developed several decades ago called stamping or die cut circuitry. The principle is simple. Loosely bonded copper-polyester laminate is passed into a large die cutting press. The die cut tool defines the conductor pattern by ''kiss'' cutting the copper. The die may have a heated insert that heat bonds the circuit portion of the copper while the undesired metal remains only tacked. The excess copper is mechanically stripped away to leave the finished circuit behind. The process is typically done in roll-to-roll format and is highly cost-effective. There is no wet process step and virtually no pollution associated with the circuit patterning step. The waste copper is easily recycled since it is almost pure, uncontaminated metal.

There are two major disadvantages. First, only low density circuits can be manufactured with very wide (typically .060″) traces. A second problem is expensive tooling with long schedules for building. Changes are difficult to make without scraping out the tool, although segmented dies and inserts have been used. This process has been used to make the large automotive instrument cluster circuits found behind the dashboards of most automobiles. The high tooling cost and inability to do fine lines, prompted Packard Electric (GM) to phase out this process during the late 1980's. However, the method is still used in Europe and Japan.

7.5 INTRICACIES OF FLEX CIRCUIT MANUFACTURING

While the previous discussion laid out basic steps in circuit manufacturing, it is interesting to discuss, in more detail, some of the intricacies of flex circuit manufacturing. Up until now, most of the basic steps described could just as well describe a rigid PC board fabrication process. As mentioned before PCB's and flex are very similar in the fabrication steps, but the following will highlight some of the key differences.

7.5.1 Material Handling

PC boards are processed in discrete panels. They are usually relatively rigid and do not offer any particularly difficult challenges in handling, such as pickup, edge registration, stacking, and conveyorizing. Flex, on the other hand, is thin and flimsy. Sheets of flex panels do not easily transport over conveyor rollers as do PC boards. Handling flex roll-to-roll gets around some of those problems but creates another set of problems to resolve.

In the case where flex is built in sheet form, it may be impossible to simply lay a sheet on a roller conveyer and have it traverse through an etcher. Such an attempt will more than likely result in the sheet never making it to the other side. Normally, the sheet of flex is "rigidized" by mounting it on a carrier frame or adhering it to some other rigid surface. Since flex is very susceptible to denting or creasing, special care must be taken in every operation to ensure damage free material. The typical wet chemistry lines, use conveyor rollers that resemble small pizza wheels spaced at several inch intervals to allow for adequate spray reaching the panel. The same type of equipment can be used for flex, but the wheels usually are softer and smaller and spaced more densely. Extremely high spray pressure can cause creases in flimsy material, so these machines may have to be configured differently.

In the case where flex is built roll-to-roll, each process step has the following characteristics.

1. Unwind station and rewind station. It is critical that these stations guide the web for good tracking and maintain proper tension for minimal distortion.

2. Proper tensioning. In order for the web to rewind in a straight consistent manner, a fair amount of tension must be put on the web ahead of the rewind coil. Yet, this tension must be isolated from the web prior to the unwind to avoid having this tensioning travel through the particular process. In many cases, it is desirable to have zero tension on the web as it goes through the process. Material handling is accomplished with a belt conveyor or some other transport.

3. Speed control. With a continuous roll moving through a process, controlling speed is not as simple as controlling the rate of the conveyor system through the process. The tensioning of the web may dramatically effect the speed. It is possible to have the rewind coil control the speed by "dragging" the web through the process. This alone may cause too much tension on the web and result in distortion. In cases where several chambers or processes are strung in line, the system must be configured to allow for the web to go through each step at the same speed, as it is impossible for one end of a web to travel faster than the other.

7.5.2 Dimensional Stability

Because of the way flex materials are laminated together and with the types of films used, a sheet or roll of flex material will tend to distort or change dimensionally through the process more than a rigid PCB. Much of the change is a result of relieving laminated in stress, which is inherent in a roll-to-roll laminating process. As a result, certain compensations must be made to allow for this distortion to occur.

7.5.3 Registration

A rigid PCB panel may be easily registered to a position relative to a tool by edge guides pushing the board against stops. This will not work well with a sheet of flex and is not applicable to a roll of flex. The following are some methods for registering flex material to a tool.

1. Mechanical pin registration. Prepunched holes in the flex panel can be used to pin the panel to a precise location. This method can be used for either sheets or a continuous roll. It is most commonly used with manual placement.

2. Mechanical web tracking system (for roll-to-roll). This is similar to mechanical pin registration, however, the entire roll has continuous tracking holes or slots cut along the side in the configuration similar to computer printer paper. The process requiring registration would use a tractor feed mechanism to handle the web.

3. Optical registration. Fiber optics and video camera technology are commonly used today in screen printing, photoimaging, die cutting, and electrical testing where precise registration of the web to a tool (phototool, die, or fixture) is required. This technology can be used for r/r and sheet flexible circuitry. There appear to be many methods for using this technology. One method is to mechanically position the web near the target area. For r/r flex, precision nip rollers could advance the web a set distance. If the above was done properly, the fiducial mark or prepunched registration mark is in the field of view of the camera, whereby the vision system takes over and guides the tool or the web through a series of movements and checks, until the pattern is determined to be within acceptable registration to the tool. This iterative process normally takes less than one second. Optical registration systems have shown to improve registration from $\pm.007''$ to less than $\pm.003''$.

Typically, the higher the accuracy built into the system, the slower the process cycle.

SUMMARY

The basic process steps of manufacturing flex circuits are very similar to the basic steps of PCB fabrication. In fact, often the same machines are used with slight modifications. The main differences come in the material handling and dimensional stability through the process and differences in the thickness of the materials. Since flexible circuits flex, they must be thin and pliable and PC boards are not, the materials used are usually very different. Creases and dents are a large concern in the manufacturing of flex because of the applications that are ideally suited for flex and the diecutability of flex. There are often complex and unusual shapes that flex circuits are cut into which adds some additional challenge in the flex fabrication process.

References

1. Frecska, T. Screen Printed Photoresists, *SITE*, pp. 22–25, Feb. 1988.
2. Gilleo, K. Screen Printing vs. Photoimaging, *SITE*, pp. 24–27, June 1988.

8

Integrated Features

Ken Gilleo

Poly-Flex Circuits

8.1 INTRODUCTION

Flexible circuitry allows many special features to be incorporated directly into the circuit. The unique properties of flex make designing and fabricating unique features practical. Important properties and characteristics include extreme thinness, compliancy, resilience, and conformability. Processing attributes of the dielectric base film like ease of piercing, blanking, and etching, are also important in creating integrated features and functions. The most important single property of flex, in terms of feature creation, is the freedom to easily bend, roll, and twist the circuit into different shapes in multiple planes. Flex is put to excellent use in applications where it serves as its own cable and connector. Integrated cabling is one of the most effective and widely used features of flexible circuitry. However, many other important integrated functions can be added, alone or in combinations, to produce unique solutions to problems that have not been solved by any other single technology. We will explore the most common and suggest that many are still to be discovered, designed, and fabricated.

8.2 BUILT-IN CABLES

Flexible circuitry can be designed into the most versatile and reliable interconnect cables in the electronics industry. Conventional methods of interconnecting all rely on joints and junctions, cables, and plugs. Each individual joint is a source of signal degradation and potential failure. High density connection cables contain dozens or even hundreds of junctions. Since a connector must be attached to each cable and circuit board, there really are four junctions per line. Where several hardboards are connected together, hundreds of failure-prone junctions will be created—unnecessarily! Flex eliminates the need for all of these connection points. The connectors are eliminated. This is the integrated cable concept. It solves the problems of junction unreliability, signal degradation, and miswiring. Weight and size reduction are additional benefits. Clean and clever single flex designs replace a conglomeration of bits and pieces.

We will briefly review the flexible circuit manufacturing process to comprehend how the continuous circuit-cable product is feasible. Copper laminate or adhesiveless clad is made with a continuous conductive layer. Copper conductor can range from microns to mils thick. The copper is chosen to provide the right current capacity, flexural fatigue resistance, and other necessary properties. Polymer Thick Film (PTF) conductors can also be used if the current demands are not too high. Conductors can be arrayed in simple parallel lines or more complex arrangements, depending on the application. The design can start with large 50-mil wide lines, for example, and taper down to a very fine pitch. The key is this: a continuous conductor arrangement is used so that all junctions and joints are completely eliminated.

The simplest flex cable is a single-sided circuit with parallel conductors arranged to terminate flush at the two ends. This type of cable can be used to join two rigid boards together, using solder joints or connectors. However, the rigid boards can be replaced by flex circuits. In this situation, the flex cable and associated circuits are designed as a single integrated system. In fact, a series of circuits and cables can be designed and built as single or multilayer structures. A number of cabling elements can emanate from a central circuit to produce "flying tails." Figure 8.1 shows several designs.

The addition of cables into a design is a simple matter of laying out conductors and cut lines. The same guidelines apply as for any dynamic flex. Flexing, or dynamic motion planes, should be oriented so that the fold motion is at right angles to conductors. Circuit component zones, between the flexing areas, can be selectively rigidized with a variety of stiffener materials. Figure 8.2 shows the proper angle orientation.

8.3 INTEGRATED CONNECTORS

We have just reviewed the integrated cabling concept, which is used to mate together circuit zones that must move or be placed in different dimensional planes in relation to one another. But flex must be connected to other circuits or electronic devices. Flexible circuit *is* a major interconnect technology, meaning that connections are an important area of design and development. A myriad of plugs, jacks, connectors, and other terminations are available as discrete

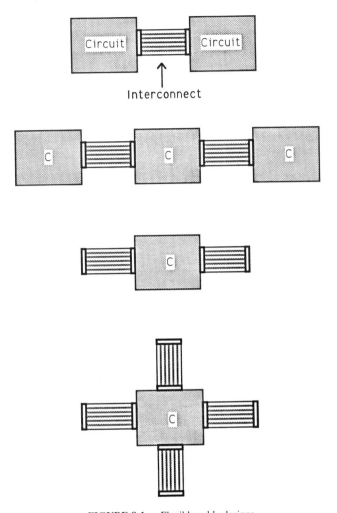

FIGURE 8.1   Flexible cable designs

products. These are covered in other chapters. We will continue with the theme of integrated features and move down the cable to the termination. Eliminating the discrete connector device can reduce weight, area, thickness, reliability problems, and cost. Performance can also be increased by integrating the connector. Excellent impedance control can be achieved with terminations that are a part of the conductor. Very high density is also possible, higher than by any other means.

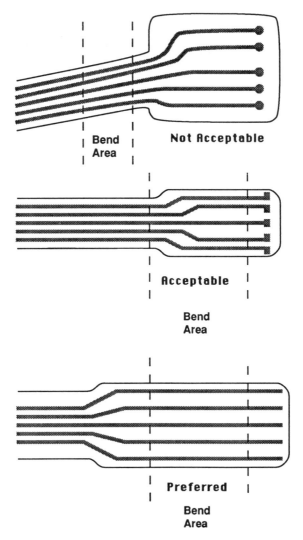

FIGURE 8.2   Dynamic flexing motion

8.3.1 Bared Conductor Termination

The simplest integrated connector is a bare conductor pattern. We can simply design the circuit pattern to accommodate the pitch required by the connecting device, analogous to the hardboard edge card concept. The cover layer or cover coat is brought to within a specific distance of the edge leaving the required trace exposure. Both copper and Polymer Thick Film (PTF) conductors can be used. A common practice in the membrane switch industry is to connect the PTF silver or carbon in conductors directly to hardboard circuits and discrete devices. A simple pressure loading arrangement is sufficient for producing a reliable mechanical interface. Bare copper conductors can be used in the same way, but copper's tarnishing tendency requires special connection designs. A protective finish is commonly used on copper terminations to eliminate the tarnishing problem. Table 8.1 compares common finishes.

The finishes can serve two important functions. First and most important, the finish should provide a good electrical contact surface. Bare copper is subject to oxidation to the extent of becoming nonconductive. Finishes greatly reduce this tendency. A second function, especially for the PTF inks, is a softer, more compliant surface that can make a more complete surface contact with the mating structure. Carbon ink, for example, is often printed over solder plate when the mating surface is a glass LCD. The carbon ink, made with a polymer binder, cold flows under the pressure produced by a clamping device. Solder plate alone is relatively hard and irregular and cannot produce the 100% contact of the PTF ink. PTF ink, printed over solder plate, overcomes the tendency toward intermittent opens with finer pitch connections when only solder is used. Solder plate *should not be connected directly to LCD's.* The information for this statement was gained at the expense of a massive field failure.

The flex circuit, with the appropriate finish, must be connected to the intended hardboard, component, or other electronic product. This is usually accomplished by means of a mechanical pressure device that forces the flexible circuit against a mating conductor arrangement on the device. Various spring clamps are commercially available. A simple spring clamp is shown in Figure 8.3. The device, bonded to the hardboard, simply exerts a force to press the flexible circuit against the corresponding conductor pattern on the rigid board.

TABLE 8.1   Protective Finishes for Metal

Finish	Typical Thickness	Reliability
Solder plate	0.25–1.00 mil	moderate
Carbon ink	0.4–0.8 mil	low–medium
Imidazole[a]	monolayer[a]	low–medium
Benzotriazole	monolayer[a]	low–medium
Rosin coating	.01–.1 mil	medium
Carbon ink over solder	same as listed	high–moderate
Immersion Gold	0.1–0.5 micron	moderate gold over
Nickel	1–4 microns	very high
PTF carbon over silver	0.4–0.8 mils each layer	moderate[b]

[a]Chemical treatment conversion coating.
[b]Silver without a carbon overlayer can suffer from silver migration, which produces shorting if moisture is not excluded from the junction, and DC current is present.

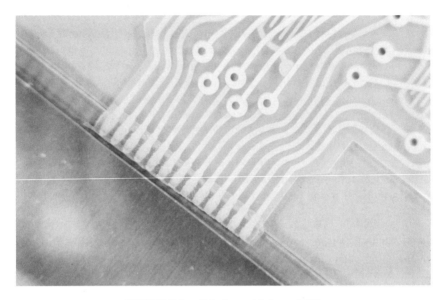

FIGURE 8.3   Calculator with flex-to-LCD

Other clamps use a lever mechanism to produce force. Even a bezel or a housing can be used to supply force, regardless of the source of the force: the concept for this simplest form of integrated connector is the same—compress the flex against an opposing conductor pattern.

Bare copper direct connections are used in high volume applications with designs that take into account copper's tarnishing tendency. The largest application is automotive instrument cluster circuits, which were introduced in the 1960's to replace point-to-point wiring. Copper polyester flexible circuitry has become the dominant dash board interconnect for the world wide automotive industry. These copper-polyester circuits are used today to supply power and signals to various dials and gauges. Power is also supplied to warning lights and general instrument lighting. A number of interconnection features have been integrated into the cluster circuit. Virtually all of these circuits are made with copper treated with antioxidants. This antioxidant treatment is sufficient for high current and long-term reliability because of the wiping action designed into these connection points.

Light bulb sockets are created as a simple annular ring in the circuit. Figure 8.4 shows a common configuration.

Split rings can be used for power and ground. The bulb holders can be designed to mate against the power and ground traces of the flexible circuit, which

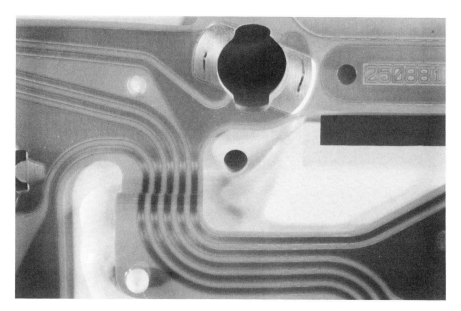

FIGURE 8.4   Instrument cluster bulb socket. (Courtesy of Sheldahl, Inc.)

is arranged over a plastic instrument housing made possible by the conformal properties of flex.

Often, a main bus connection structure is designed into the flex, which brings power and signal from the fuse buss. Designs involve a cutout around the connector array that allows the circuit to be pressed against the fuse buss. A common design is shown in Figure 8.5.

The next level of flexible integrated connection is the semi-permanent type. The same basic configuration, bared conductors, can be used. The difference is that the flexible circuit or mating device provides material, which forms an electromechanical connection. The flex circuit can be solder coated to become the total source of solder. A hot bar soldering technique can be used to form the solder reflow bond between the circuit and mating device.

A number of flexible circuit edge designs have been conceived over the years. The presence of the base film under the entire circuit edge makes soldering more difficult and tends to induce solder bridging. One solution is to cut out the dielectric between conductors. This also makes inspection easier as well as cleaning. The process of removing dielectric film can be accomplished with a mechanical blanking process. If dimensions are not too small, chemical milling, with high etch factor, or laser machining can be used. Figure 8.6 shows the interdigitated termination concept.

A more advanced design involves the complete removal or elimination of

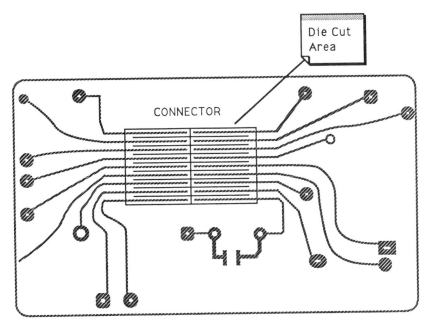

FIGURE 8.5   Instrument cluster circuit integrated connector

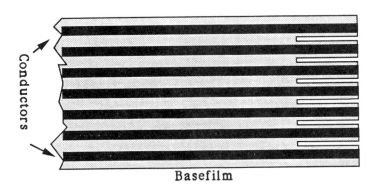

FIGURE 8.6   Interdigited end connection. (Sumitomo Ford Circuit)

dielectric in the conductor termination area. Bare copper or plating-over-copper conductors can be beam leaded just like TAB. The unsupported conductors can now be more easily bonded to an electronic device or hardboard. This design reduces pitch considerably. In fact, the product can be made using TAB pitch criterion. Circuits have been made with a pitch of 10 mils, without really pushing the limits of technology. A common practice, especially with fine pitch unsupported leads, is to leave a copper tie bar in place to hold the cantilevered conductors correctly oriented. The tie bar is removed prior to bonding or as a sequence in the bonding step. Figure 8.7 shows this design concept.

The cantilevered beam design also permits thermal compression bonding to be used for very fine pitch connections. There is even a higher performance level of design where special connector features are formed and integrated into the flex circuit. Two popular technologies are Sculptured® circuits, by Advanced Circuit Technology, and Gold Dot® by Hughes. These systems are available from licensees.

8.3.2 Sculptured Connectors

The Sculptured® circuit process is a form of step etching practiced by the chemical milling industry and others. The basic process involves starting with extra thick copper laminate, up to 10 mils or more of copper, and etching down all but the termination zones. The circuit and connector patterns are then photoetched in the usual manner. The result is a sturdy integrated end connector

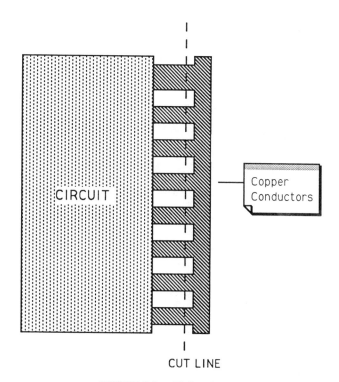

FIGURE 8.7   Tie bar diagram

that can be plugged into a socket or used in any other connection format that is applicable. The key is that the circuit and the connector are fashioned from a single piece of copper. This, of course, provides very high reliability and signal purity at maximum density and minimum size and weight. Figure 8.8 shows a sculptured end connector.

8.3.3 Gold Dot

The Gold Dot® process provides the highest density per area with the ultimate in reliability for a reconnectable system. Higher cost is the penalty, however. The method involves plating up small gold bumps or dots over the entire connection contact area. This means that the connection points are not limited to the edge only. The gold bump height, shape, and hardness are controlled to provide an extremely reliable, very low ohmic interconnection capable of thousands of connect-disconnect cycles. Specific clamping pressures and designs are required. The process is used under license by Hughes and has been mostly used for military applications.

Various modifications of the concept have appeared primarily in the commercial electronics sector. A reportedly lower cost version involves first plating up copper posts followed by gold plating. Developers claim that the copper post provides more dimensional stability at a substantially reduced gold cost. It is not yet clear if any of these modifications are available and if they are covered by patents.

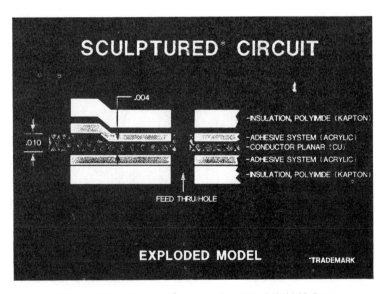

FIGURE 8.8   Sculptured® circuit. (Courtesy of Sheldahl, Inc.)

8.3.4 Dimples

A mechanical embossing process was developed in the late 1970's by Sheldahl and Hewlett Packard to solve an interconnect problem for the Ink Jet® printer. The concept involves forming small domes or "dimples" on the mating surface of flex circuit conductors. The goal was to produce a remarkably moderate- to high-density interconnect at reasonable cost. The tiny embossed bumps are usually gold plated to ensure an oxide-free mating surface.

The advantage of this process is that the connecting dimples are produced by forming the copper metal and, thus avoiding electroplating. Deficiencies include limited density because dimple diameter and height are limited by the type and thickness of copper, adhesive, and base film. Another limitation is that of maximum contact force. The dimples can collapse if too great a force is applied. Improvements have been made both in the forming process and in materials. The dimple connector should be considered where lower cost, remakable connections are required.

8.3.5 BetaPhase

A self-closing connector has been developed, using flexible circuitry as the principle component. Close-on-demand action is provided, using a memory metal alloy spring. Memory alloys, first discovered by the Naval Ordnance Lab several decades ago, undergo a major phase transition at a specific temperature. Application of heat transforms the alloy into a spring. Heat is provided by a heating element in the circuit. The end result is a pressure connector end, which closes when current is applied to the heater element. BetaPhase Corporation offers this patented concept.

8.4 DIRECT CHIP INTERCONNECT

8.4.1 Integrated Beam Lead Versus TAB

Tape Automated Bonding (TAB), is a technology that utilizes a specialized flexible circuit product as a chip carrier. Modern TAB consists of a thin flexible dielectric film with a conductor patterned in an array that is designed to be bonded to connection pads on an Integrated Circuit (IC). The portion of the conductor intended for connection to the chip is cantilevered over an opening in the dielectric. These beam leads, called "fingers," are compression bonded to IC pads by means of a heated pressure tool, called a thermode. The opening in the dielectric, referred to as a "window," allows the thermode to directly apply heat and pressure to the TAB beam leads. Although the window design is not absolutely necessary, it allows the bonding thermode to make direct contact with the TAB bonding lead resulting in a faster, higher quality bond that can be easily inspected. Figure 8.9 shows a diagram of TAB, while Figure 8.10 is a photograph of today's TAB.

The mating of the IC to the TAB conductor array requires that the right mechanical geometry and metallization be available. Gold-to-gold or gold-to-tin eutectic bonding are the most popular, although copper-to-copper bonding is in use today. The IC is usually gold plated while the TAB "fingers" are either gold-plated or tin-plated. In order to connect the end of the TAB fingers to the IC, a raised metal structure appropriately called a bump is usually provided. The bump can be formed on the TAB conductor, but placement on the

TAB FRAME EXCISED FROM TAPE

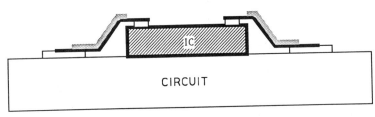

TAB ASSEMBLED TO CIRCUIT

FIGURE 8.9   TAB diagram

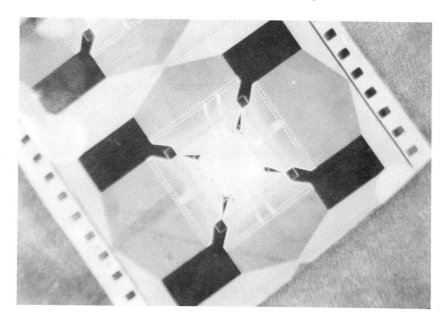

FIGURE 8.10   Photo of TAB

IC pad is becoming increasingly more common. The bump can be formed on the IC with more precision and in a way that seals the IC.

The purpose of the TAB carrier is primarily to provide a simplified interconnect between the very fine pitch IC and the wider pitch printed circuit. The IC fabrication process is capable of producing dimensions that are less than 1/100th the size that can be produced by printed circuit techniques. TAB bridges the IC to the real world. The TAB configuration provides a pre-configured conductor array, called inner leads, which exactly match the IC pad array. The outer lead TAB arrangement matches connection pads on the printed circuit. In one sense, TAB is an array of "wires" that can be conveniently held in alignment with the devices to be interconnected. A 4-mil pitch device is "wired" to a 20-mil or even 50-mil pitch circuit. We will view TAB as a small, fine line flexible circuit used to replace point-to-point wire bonding. Early flex began as a simple wire replacement but for large connection sites with very wide pitch.

TAB has a few more features beyond serving as a fine pitch wire array. It becomes the carrier and the package for the IC. We can view "loaded" TAB, the product that has a bonded IC attached, as a component in a package. A reel of TAB can be loaded with chips and supplied to an electronic assembler. The IC's can easily be tested while in the reel before committing them to the circuit. Test pads are usually designed into the TAB circuit. Large IC's, especially with a high I/O count, are often packaged as individual circuits in frames to avoid bending stresses. Now let's look at the bond-to-circuit, or Outer Lead Bonding (OLB) process.

There are three basic TAB constructions, each made with a completely different process. Single layer TAB has only a conductor layer and no dielectric. The tape is a continuous reel of copper that is bussed together to maintain mechanical continuity. The IC cannot be tested, while the single layer TAB is in reel form since all "wires" are shorted together. The individual circuit must be cut out after the IC is bonded so that the array will not fall apart. Two-layer TAB consists of a conductor foil bonded directly to the dielectric without an adhesive layer. In flex terminology, the product is simply an adhesiveless clad. The dielectric must be removed after the circuit features are formed. The most common TAB product is three-layer, conductor/adhesive/dielectric, which is identical in construction to conventional flexible circuitry. This product can be made by pre-punching adhesive-coated dielectric before laminating to copper foil.

Whatever the construction, the bonded IC is usually protected by plastic in some form. The one-layer TAB is commonly injection molded after IC attachment to produce a product called TapePak™, by National Semiconductor. The 2- and 3-layer TAB product is typically covered with liquid resin or monomer around the IC and ILB area. This potting or globbing process provides adequate protection for most, but not all, applications. The IC's are usually passivated with an inorganic coating, and the glob just adds more protection.

The Chip-on-TAB must first be cut out of the framework array that provided continuity for the reel carrier feature and test pad array. The flex circuit industry calls the cutting out of individual circuits, blanking. The TAB group likes the term "excising." We will not need to debate the terminology since we will eliminate the process in the next paragraph.

The excised or blanked out miniature flexible circuit is ultimately bonded to a larger printed circuit by compression bonding or fine pitch soldering. Polymer bonding techniques are also coming into use. The blanking and bonding processes are often merged into a single operation. Once all of these steps are accomplished, the result is a reliable, lightweight and low profile flexible circuit assembly provided that the base circuit is flex. Flex-on-flex seems a bit redundant while flex-on-hardboard seems like a good marriage of two technologies. We can question the value versus the negatives for the flex-on-flex scheme.

Pretestability provided by TAB is a useful feature in many cases, but progress in wafer-level testing continues to reduce the value of such an attribute. Memory chips, for example, are efficiently tested in wafer form. Setting aside the pretestability issue, let us examine the logic of fabricating a small flexible circuit and then bonding it to a larger one. Because both circuits can be made of the exact same materials, why not simply integrate the smaller flex into the larger? Theoretically, we can make a flex circuit with fine line conductor arrays and construct windows in the base film. We can call this product TAB-Flex, TAB-Featured Flex, Beam Lead Flex (BL-Flex) or some other descriptive name. Because the tape feature will ordinarily vanish on integrating the chip bonding array into flex, we will avoid the TAB association. The TAB industry also uses terminology that has different definitions than the larger flexible circuit industry, and we want to avoid this confusion.

The Integrated Beam Lead flexible circuit has most of the features that are valued in TAB. Pretestability has been given up, however, but valuable enhancements have been added. The most striking difference between BL-Flex and TAB is that we have totally eliminated Outer Lead Bonding (OLB). The IC is connected directly to the flexible circuit. We have a true direct chip interconnect design: there is only one level of connection. TAB has twice as many connections and twice the potential for junction failure. Impedance is superior and signal lines are shorter and the size of the total circuit is now at its theoretical minimum. Is the scheme workable and cost effective?

An interesting historical note is that the integrated beam lead flexible circuit was devised long before TAB. In fact, TAB is really a spin-off of flexible circuit technology. MIT[1] and others,[2] produced beam lead circuits made of copper Kapton® flexible circuitry. These designs, produced during the 1960's, had the unsupported beam lead feature and were made of copper Kapton® flexible circuits. The common designs of that period had dozens of IC bonding sites and were really early multichip modules. Figure 8.11 shows a comparison diagram between Beam Lead Flex and TAB. Figure 8.12 shows a photo one of the earlier circuits complete with bonded IC's and surface mount components.

During the later 1960's, reels of beam lead flex constructions began to appear. GE was one of the early pioneers with their Minimod tape, which was patented in 1972.[3] Later, Honeywell coined the name, Tape Automated Bonding, which grew into the popular acronym, TAB.[4] Even though GE's Minimod is considered the beginning of TAB, similar designs were already known. Perhaps the earliest flexible tape chip carrier was invented by Hugle Industries. Mrs. Francis Hugle received a patent in 1967 for a chip carrier made of copper on flexible dielectric.[5] The Hugle invention predated the GE idea by several years and with several novel features. One was the idea of step etching the copper leads to form a bonding bump. Another was round sprocket holes instead of the more difficult rectangular ones and a third was the use of low cost polyester Mylar® instead of Kapton®. Today, millions of polyester TAB circuits are produced in Japan each month. Perhaps the round sprocket holes will be reinvented soon. Chapter 9 on Assembly discusses TAB and other Chip-On-Board technologies.

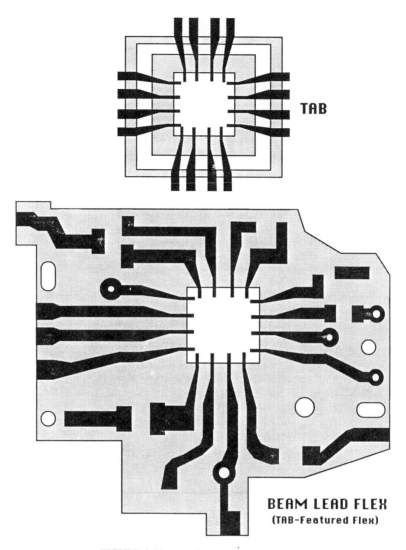

FIGURE 8.11   BL-flex versus TAB diagram

8.4.2 Plated Posts

Several plated-up post interlayer connection processes have emerged during the late 1980's as an alternate to the more common plated-through hole connection scheme. The rational behind the plated post strategy, actually one of the earliest multilayer connection methods, is that a solid post will be better able to withstand thermomechanical stresses. But there are some specific chip interconnect

FIGURE 8.12   Photo of BL-flex. (Courtesy of Sheldahl, Inc.)

features that can be created with the plate-up process. Posts for bonding bare die can be produced at any layer level. One concept, suggested by Unistructure, Inc., a maker of plated post circuitry, is to plate bonding posts for IC's. Cavities within the multilayer circuit structure can be produced for nesting the chip. Subsequent layers can be used to constrain the chip and maintain contact force. Although the plated posts for the chip bonding concept is being applied to hardboard, the compliant nature of flex provides some strain relief for thermomechanical forces. However, no reliability data or junction performance data has been published.

8.4.3 Flip Chip

Flip Chip, or Controlled Collapse Chip Connection (C4), has been used on ceramic substrate because that circuit material closely matches the Temperature Coefficient of Expansion (TCE) of silicon IC dies.[6] Bare IC dies are soldered directly to the circuit board in a precision reflow process. The solder as provided by the solder ''bumps'' on the chip, are created in a combination thin film and chemical plating sequence. The microscale of this technology requires a very close thermal match to avoid breaking or weakening the small solder junctions that are closely spaced together. Organic-based circuits typically have too high a thermal expansion to be used here.

Flexible circuitry offers an alternative. Two properties of flex make it viable for flip chip and other Chip-On-Board (COB) packageless concepts. One is the relatively low TCE value of the polyimides used for flex. Newer polyimides have TCE values that can match those of silicon, although products are now being offered at about 17–18 ppm to match the expansion of copper. Copolymer polyimides are being sampled with TCE values around 3–4 ppm comparable to silicon components.

The second important attribute of flex is compliancy. The base film can bend out of plane to accommodate thermal expansion during heating as well as elongate under stress. As will be seen in Chapter 9 on Assembly, flex provides substantial strain relief for Surface Mounted Components (SMC's), thus eliminating solder joint cracking due to thermomechanical reasons. The same intrinsic strain relief principle can be applied to flip chips, which are really the ultimate in down-sized SMC's. However, the relative size difference between SMC's and flip chips requires that either a thinner base film be used or that it be matched more closely to the TCE of the chip. The flip chip is so small and thin, that the flex base film acts like a thick, rigid board.

Flip Chip-On-Flex is being evaluated for several applications, including automotive. It is also very likely that IBM did much undisclosed work with flip chips on flex since they have had both technologies for decades. The outlook for ultrathin flex and low TCE flex in Direct Chip Interconnect (DCI) applications is very bright. This potentially excellent Multi-Chip Module (MCM) media can provide very robust, high density packages. The added ability to produce unsupported cantilevered beams (TAB-Featured Flex), can allow both flip chip and Integrated TAB bonding on the same circuit, providing two level-1 (one connection between chip and circuit) approaches. Chapter 9 discusses these and other concepts for assembly.

8.5 INTEGRATED RESISTORS

Several technologies can be used to combine resistors into a flexible circuit configuration. Thin film, thick film, and electroplating have all been used to incorporate resistors directly into a flex circuit. Resistors can be created on outer conductor patterns or within a multilayer structure.

8.5.1 Thick Film Resistors

Polymer Thick Film printed resistors have found successful applications in the flexible circuit environment. The two important characteristics of PTF resistors for flexible circuit applications are relatively low temperature processing and flexibility. PTF resistor ink consists of binder, such as polyester, carbon particles, and solvent. Most inks are the solvent evaporation type, which allows them to be quickly dried at temperatures as low as 100°C. Available resistance values range from about 10 ohms to over 100K ohms per square. Resistors are screen printed onto various substrates, especially polyester and polyimide. Several design strategies can be used to create the connection to the resistor ends.

PTF resistors are popular for polyester-based circuits made with PTF conductive ink, such as silver. Resistors can be printed first, followed by printing the conductors. The conductor ink is allowed to overlay the ends of the resistors so that a good junction is made within the tolerances of printing. Some prefer to print resistors first and conductors next so that the resistor lays flat on the substrate. The resistor ink also gets a free bake when the conductor ink is dried. This helps produce a more stable resistor. The disadvantage of this sequence is that resistors cannot be easily tested until the conductor ink is printed and dried.

The majority of flex circuit manufacturers print the resistor last so that resistance can be checked on line and adjusted accordingly. Since PTF conductors have a typical thickness of about 0.4 to 0.8 mils, the resistor junction is reliable. The edge of the conductors tends to be rounded, which creates a low stress resistor junction. PTF resistors are most commonly used to drop voltage and to decode a matrix. LED applications, for example, can use PTF resistors to drop voltage to the required value. Because silver conductors add some resistance, the dropping resistors can be designed to compensate. Resistance values are determined by Ohm's Law. A simple adjustment in the aspect ratio, length/width, provides the right resistance. See Eq. 8-1—Resistance

$$\text{Resistance} = \frac{(\text{Volume resistivity of ink}) \times \text{length}}{\text{width} \times \text{thickness}} \tag{8-1}$$

Volume resistivity should be in ohms/square/mil, if thickness is in mils. The length is measured between the conductor junctions as shown in Figure 8.13. This is a close approximation since the minor termination effects are not included.

Resistance Value Adjustment.   PTF resistors can be laser trimmed on both polyester and polyimide but cost constraints often do not allow this luxury. The more common practice is to print to value. The nearest value resistor ink is used with the aspect ratio that is calculated to give the desired value. General guide-

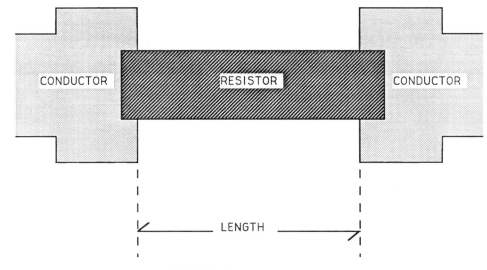

FIGURE 8.13   Resistor termination

lines are to use aspect ratios ranging from 2 : 1 up to 5 : 1. Higher ratios are used for digitizer resistor lines, where a very long resistor is printed over connector traces. When a specific switch in the array is pushed, a unique resistance value results, and the switch location is then electronically decoded. Digitizer pads have also been designed where the active decoding area is one large resistor. When a specific location on the pad is pressed, a unique resistor value for X and Y is produced to allow the location to be determined with an analog decoder. A typical resistor decoded circuit is shown in Figure 8.14.

The resistor value can be adjusted on the run by changing parameters on the screen printer. Squeegee pressure and speed changes will change the amount of ink deposited. As seen from Eq. 8-1, the thickness affects the resistor value. PTF resistors are also used with copper flexible circuitry. The resistor must, of course, be printed after the copper circuit pattern is created. The problem in printing resistors onto copper circuitry is that the conductors are typically 35 microns thick (1.4 mils) and have a sharp edge profile. PTF resistor ink can be printed up onto carbon traces, but the junction represents a high stress point. The problem can be overcome by printing PTF conductor ink onto the copper termination to produce a PTF termination. Silver ink can be used that has a much higher degree of flexibility than carbon, due primarily to the platelet shape of the silver compared to the spheroidal carbon.

8.5.2 Etchable Resistor Films

Flexible material is available, consisting of metallic resistor film and conductive foil on a flexible dielectric base. Special etchants are used to produce resistor

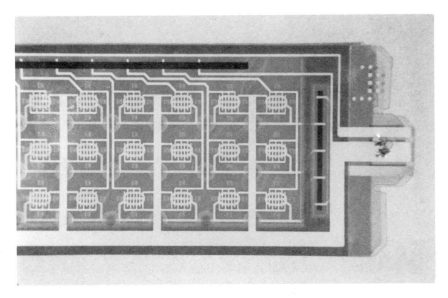

FIGURE 8.14   PTF resistor decoder

arrays. Omega Corporation supplies resistor sheets and holds patents in this area. Resistor layer sheets, once fabricated, can be used as inner layers in multilayer circuit constructions.

8.5.3 Thin Film Resistors

Vacuum deposited metals can be applied to flexible substrate directly to create precision resistors. One very successful application has been for the automotive industry. Airflow sensors have been manufactured, using thin film technology on polyimide. Air movement over the surface of the resistor cools the electrically heated element. Resistor materials are chosen for a high Temperature Coefficient of Resistance (TCR). Increased airflow lowers the resistor temperature, which subsequently changes the resistance value. Alloys, such as nichrome have the right resistance range and a high TCR.

8.5.4 Electroplated Resistors

Nickel alloys, containing phosphorus, have relatively high electrical resistance. Processes are available for electroplating these alloys onto dielectric surfaces. However, most of this work has been applied to rigid substrates.

SUMMARY

Flexible circuitry is the most integrated and potentially integrateable interconnect medium in use today. Many products such as TAB, are really flex circuits in disguise. The thinness, ability to be bent and repeatedly flexed, and the ease of machine and shaping, make flex the "most-things-to-most-designers" circuit, interconnect and packaging technology. Flex will help push the frontiers of volumetric density even further, and density demands will push the technologist to invent even more flex products. We wish the creative reader much success in developing a new flex circuit integration feature.

References

1. Gilleo, K. Substrate Windows, *Electronics*, pp. 52–54, August 19, 1968.
2. Anonymous. IC Ribbons, *Electronics*, pp. 36–38, July 24, 1967.
3. Anonymous. GE's Minimod, *Electronics*, p. 33, Dec. 7, 1970.
4. Scrupski, S. IC's On Film Lend Themselves To Automated Handling By Manufacturer And User, Too, *Electronics*, pp. 44–48, Feb. 1, 1971.
5. Hugle, F. Automated Packaging of Semiconductors, U.S. Patent 3,440,027, April 22, 1967.
6. Miller, L.F. Controlled Collapse Reflow Chip Joining, *IBM Journal of Research and Development*, pp. 239–250, May 1969.

9

Assembly

Ken Gilleo
Poly-Flex Circuits

and

Scott Lindblad
Sheldahl, Inc.

9.1 INTRODUCTION

Imagine a circuit that can be populated as a single-sided assembly, folded into any configuration, have "tails," or pendant interconnect cables, which flex hundreds of millions of times, eliminate solder joint fatigue, and be more cost competitive than hardboards on a total assembly basis. The technology is here and it is called flexible circuit assemblies.

The thinness, conformability, and compliancy of the flexible circuit makes it a truly unique product for component assembly. The compliant nature of the very thin base dielectric produces a nearly strain-free assembly even when leadless electronic components are used. This one feature alone solves the solder joint thermomechanical fatigue problem, which plagues most rigid board constructions. This means that flex and Surface Mount Technology (SMT) come together to produce an ideal synergy. Each technology supports and enhances the other. SMT-Flex assemblies can withstand thousands of thermal cycles. The strain-relief mechanism is superior to matched TCE concepts since localized component heating does not cause problems for SMT-Flex.

Surface Mount on flex has many dynamic advantages—the most pronounced being the reduction of thermomechanical stress in thermally active environments. Thermal stress is one of the most serious causes of solder joint failure in the real world today. With the accelerated rate of electronic technology advancements, we are requiring electronics to survive in extremely active environments, and we rely on their performance with our lives: for example; airliners with total electronic flight control systems, automobiles with electronic steering, life support systems with electronic monitoring and control, and missiles with electronic guidance and control systems. All are controlled by electronics, and all will see thermal stress. In a thermally active environment, the more rigid the board, the sooner the assembly will fail.

Other benefits of the thinness of the dielectric base include excellent heat transfer and the ability to conform to housings and shaped backer boards. This means the flex assemblies can be used in three-dimensional configurations to maximize unit volume density. These important features of high junction reliability and assembly conformability have boosted the use of flex in a number of areas. We will explore basic design rules, assembly techniques, and actual applications in this chapter.

9.2 COMPONENT TYPE

Feed through or leaded devices have been assembled to flexible circuits for several decades. However, the thin base dielectric, unless reinforced with a backer, must be handled with care to prevent tearing of the base, delaminating of the copper trace, or stressing of the solder joint. Conservative design rules, therefore, require that a rigid backer be used when feed through components are used.

Since the early 1980's flexible circuitry has been used with SMT with great success. SMD's are ideally suited for use with flex. The resulting butt joints are the optimum type for use with a thin, flexible dielectric. In a sense, SMT is the packaging medium that flexible circuitry was waiting for. Backers or stiffeners are not required for SMT-Flex assembly although selective rigidizing may be useful for some applications. The bottom line is that flex and SMT are mutually synergistic, and SMT is highly recommended over the older feed through assembly technology. Both SMT and flex strive and achieve the same goals. Weight and size reduction and general miniaturization achieved with both technologies are enhanced when the two come together.

9.3 MATERIALS

Assembly is the most important process in terms of materials selection. Polyimide base film circuits are used primarily to withstand the thermal stress of soldering, not for end use criteria. Although a large number of potentially suitable flexible films exist, only polyimide has gained wide acceptance for solder assembly use. Many other high temperature plastics, fall short of handling the temperature extremes of molten solder. We therefore recommend the conservative design approach of specifying polyimide for solder assembly applications. Alternative materials can be explored once the design is proven and the assembly process verified.

The information in Chapter 2 clearly shows the superior thermal performance of polyimide. Kapton® is zero strength rating of about 800°C suggests that polyimide is even overkill. Attempts to produce high temperature alternates have not met with much success. However, one material, Nomex®, an aromatic polyamide, has gained some use. Nomex® is used in a non-woven form, saturated with laminating adhesive. Although Nomex® satisfies temperature requirements, its most significant limitation is in dimensional stability. Aramids are notoriously hygroscopic and have a large dimensional change associated with humidity differences. This situation is made worse by the non-woven paper construction of Nomex®, which results in large dimensional variations in the laminate. This combination of loose dimensional tolerances and humidity-influenced dimensional changes makes Nomex® of limited value for precision circuits.

So-called advanced engineering plastics have also been investigated as base films for solder assembly destined flex circuits. Materials like polyetherimide (PEI), polysulfone (PS), polyphenylene sulfone (PPS), liquid crystal polymers (LCP), and several other polymers all fall short of requirements. Most do not have enough thermal performance reserve. Although eutectic solder melts rapidly at 181°C, soldering processes can easily cause the substrate to overshoot this temperature by 30 or 40°C.

Another approach to reducing the need for polyimide is to control substrate heating by localizing soldering or solder reflow heating. Careful design of the circuit to leave a stabilizing boarder of copper around the periphery, allows low temperature base film to be used. Designs with large amounts of stabilizing copper must be used with modified soldering processes if thermoplastic bases, like polyester, are to be used successfully. It is possible to solder polyester circuits using a number of methods, including wave soldering. We recommend that first designs and prototypes be done in polyimide. Careful review of the assembly process, ramifications, design requirements, and dimensional changes during solder are necessary before considering a switch to polyester. Alternates to solder assembly are slowly making their way into the realm of accepted techniques, and this will increase the use of lower temperature base film.

9.4 FEED THROUGH ASSEMBLY

Since it is not always possible or desirable to use SMT, general guidelines are provided for feed through components. Larger, heavier components must be supported with a backer board to be used successfully with flex. A common design involves bonding a precut and predrilled rigid board to the flexible circuit. The backer board is an inactive component strictly used as a mechanical support and planarizer.

A pattern of oversized holes, corresponding to those in the flexible circuit, are drilled or punched. This backer is than bonded to the bottom of the flexible circuit. Where a single-sided circuit is used, the backer is bonded to the side without conductors. Adhesive may be applied to the circuit or backer. Both heat activated and pressure sensitive adhesive are used. The adhesive must be applied to either substrate prior to hole formation so that adhesive does not occupy the assembly area.

One interesting design concept involves the use of selectively bonded backers in a breakaway configuration. The backer can serve as a rigidizing carrier during assembly. After assembly the circuit and bonded backer is snapped out of the array. Several circuits per carrier can be used for an efficient assembly process. The selective bonding of the backer allows the final assembly to be folded into the desired 3-D shape. The unbonded backer simply falls away during the depanelizing step. Figure 9.1 shows an assembly panel before the separate circuits are broken out. Figure 9.2 shows the resulting assembly, formed into the 3-D shape required for installation into the product.

Feed through assembly is accomplished in much the same way as for hardboard. Components are pushed through from the backer side, when backer is used. The "board" is then run through a standard wave solder line. One additional requirement is that the circuit, if made with polyimide, be pre-dried to reduce moisture content. Older polyimides, like Kapton®, absorb almost 3% water—enough moisture to cause explosive outgassing during soldering to result in conductor delamination or solder voids. A prebake of 110–130°C for 10–30 minutes will usually remove enough moisture to prevent difficulties. Newer polyimides can eliminate the need for drying, since moisture absorption is less than 1% for many of these materials.

FIGURE 9.1   Ford assembly. (Courtesy of Sheldahl, Inc.)

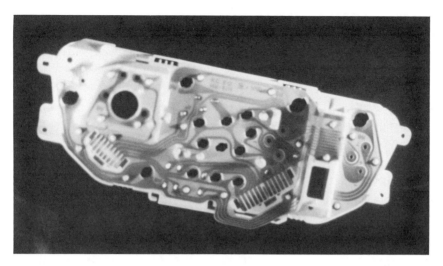

FIGURE 9.2   Ford assembly folded. (Courtesy of Sheldahl, Inc.)

9.5 SURFACE MOUNT ASSEMBLY

Although SMT and flexible circuits were meant for each other, the mating of these technologies requires patience and care. The first obstacle for assembly is handling the flexible and easily bent material. A flexible circuit must be kept relatively flat throughout most of the assembly steps. This requires a strategy to be established and implemented by the assembler. A standard rigid board assembly line can probably not be used for flex without either modification of the line or temporary rigidization of the flex.

9.5.1 Components

Components come in a wide variety of package sizes, and package types. The EIA has set the industry standard for SMT package sizes, coverlayer pull, reel sizes, and sprocket sizes. The IPC has set the standard for pad sizes, component leads, and commercial solder joint quality. The easiest component packaging medium for surface mount equipment is tape and reel. Common components found on tape and reel are resistors, capacitors, inductor coils, IC's, and diodes among many others. Component reel standard sizes are 7″ and 13″. The 13″ reels are good for high volume production or where component changeovers are to be minimized. The 7″ reels have the advantage of availability and less inventory per reel, if lower volume JIT techniques are used. Larger packages are usually placed in waffle trays or tubes. Although material handling time goes up using these packaging techniques, each package style has its advantages and disadvantages. The waffle trays do not bend the leads of the IC's and handle large IC's better than any other packaging medium. Tubes are designed for small to medium size IC's and are easier to fill on the fly. Each equipment manufacturer will have a preference as to which works better on their equipment.

9.5.2 Equipment

Equipment selection for manufacturing flexible circuit assemblies requires careful consideration before purchases are made. Most equipment manufacturers have no experience on flexible circuits and have not set aside R&D dollars for development of flex assembly processing equipment. But as the market continues to grow, equipment manufacturers will elect to play in this market. If you have never run a flexible circuit through your equipment, run to the local hardware or grocery store and pick up some copper foil. Try running a piece through the equipment to generate ideas and look at trouble spots. Flex circuits do not lay flat, which can cause vision system malfunctions. Flex circuits are not stiff enough to run through on edge rail conveyors without support. Fans in the infrared oven will blow the circuits around. High pressure aqueous cleaning can cause dents in the flex circuit, and electrical test probing can be a real treat. Each process step will require some type of modification.

A wide range of stencil equipment exists on the market from *in*-line to stand alone systems. Stenciling is one of the most critical steps in the process due to the accuracy and deposit. When stenciling on flexible circuits, a fiducial system should be used to compensate for tooling accuracy and material stretch or shrink. Some stencil manufacturers also supply input/output ports for future improvements in the stenciling equipment, which should be considered a requirement. When selecting SMT equipment one should know that standards have not been set for benchmark testing of speed, accuracy, equipment utilization, maintenance, and uptime or fiducial reading requirements. When determining the critical parameters required for a particular board, the only way to find the speed and accuracy of the equipment is to run your board and time it. Some manufacturers quoted rates at more than 50% off in speed and accuracy. Most equipment manufacturers do not include fiducial reading time, block skip time, transport time, or feeder loading into the speed numbers. Machine accuracy is also a black hole. Some equipment manufacturers quote the pick-up tool accuracy to the pad, others quote component accuracy to the pad, and the best give a 3 sigma guarantee for the component to the pad.

Management information is a very valuable asset to any automated line. Management information, at a minimum, should tell you how many components were placed, what part numbers were placed, what part numbers were dropped, equipment utilization, safety or processing errors, and SPC by tool.

9.5.3 Infrared Reflow SMT

The first step requires the application of solder paste. No precleaning or deoxidization should be necessary unless the circuits are very old, improperly packaged, or contaminated. The circuit must be held precisely flat on the stencil or screen printer. The two most common methods are to rigidize the circuit using a carrier board or to hold the circuit with a vacuum on the printer bed. Each method has its advantages, and a choice must be viewed from a total system approach. The vacuum hold down method is adequate for small circuits, but large arrays become difficult to handle for subsequent steps. The carrier board method is gaining wider acceptance, especially where flex and rigid boards are assembled on the same line.

The next step in the process is component placement. Unless the flex circuit is rigidized on a carrier, a vacuum hold down should be used on the pick and place equipment. Some assemblers have set up vacuum hold down even on shuttling type machines. A planer circuit is especially necessary to control component height and reduce bridging for fine pitch.

Flexible circuitry has a larger dimensional tolerance than rigid, which generally requires vision placement equipment. A large sheet of circuits in a "multi-up" configuration, will have dimensional variance across the sheet greater than can be tolerated with edge registration alignment strategies. This means that locating targets should be designed into every circuit as well as in corner locations. Modern pick and place equipment can then start with full sheet targets and default to the necessary level as required by the particular circuit variance. Some standard programs start by aligning to sheet edge targets, then check array cluster targets and finally, move to individual circuit targets if the sheet's dimensional variation requires this extra precision.

The loaded circuit is not ready for solder reflow. Once again, the need to keep the circuits flat becomes important and the challenge more difficult. A vacuum hold down approach, although not impossible, is certainly more difficult through an IR oven. Unrestrained circuits will temporarily curl upward during heating as the base dielectric expands at a greater rate than the copper conductors. The curl can be extreme enough to slide components off their pads, although surface tensional forces tend to hold the component leads to the pads.

The reflow process works well with a carrier board strategy. This step may be the one that can determine the handling method for all of the previous steps. We will, therefore, devote additional discussion to carrier board concepts.

A carrier board must obviously withstand the temperatures of solder reflow heating but also survive repeated heat exposures. Common hardboard materials, such as G-10 can be used. The board should have a relatively low heat capacity so that an extra heating and energy burden is not added. The ability to machine and add tooling and registration pins may also be useful. The next challenge is that of holding the flexible circuit in place. Several methods have been used successfully, and others are sure to be invented. Table 9.1 lists basic concepts with limitations and advantages. Once a successful method has been established for transporting circuits through reflow, the method should be considered for the previous steps of solder paste application and component placement. Use of a single carrier board throughout the entire assembly and test sequence has obvious benefits for reduced labor and standardization.

Once the soldering step is complete, assemblies can be handled without support for cleaning and testing. Circuits can be cleaned by vapor or spray in their flexible format. In fact, the ability of flex to bend probably facilitates cleaning under components. It may be desirable to either keep the circuit on the carrier or use a wire mesh basket for high pressure spray applications.

Other heating methods can be used. Vapor phase reflow works well with flex because of the low thermal mass presented to the system. Laser welding has also been demonstrated. None have gained wide acceptance primarily because of the cost penalties for these methods.

9.5.4 Wave Soldering

Wave soldering has been used for high volume assembly of flex in the U.S. and Japan for many years. The method allows existing wave solder equipment to be used and is a simpler method where applicable since solder paste application is eliminated. But, the process requires the added step of component adhesive attachment.

TABLE 9.1   SMT Fixturing Techniques for Flex

Hold Down Method	Disadvantages	Advantages
Tape, such as Kapton®	Labor intensive may not prevent waviness in center	Simple
Pressure sensitive low tack adhesive applied to carrier	Limited lifetime, can be difficult to remove circuits	Holds substrate very flat
Magnets	Need ferromagnetic inserts in board, high heat can destroy magnetism	Very simple, fast
Clips	Only restrains edge, spring can be heat degraded	Relatively quick to use
Vacuum	Requires substantial engineering and equipment	Very simple concept
IR transparent hold down sheet, such as quartz	Must be carefully placed	Very effective where applicable

Adhesive for component attachment must be applied to the circuit before placement. Circuits can be held on carrier boards or handled with vacuum hold down. Dots of staking or attachment adhesive can be printed, stenciled, or pneumatically dispensed. Pneumatic dispensing is excellent when a limited number of components are involved. Thermal, two-part reactive or UV curable types of adhesive can be used. Flex, with its thin, somewhat transparent base film, is unique in that UV can sometimes be applied from the back of the circuit. Polyester film circuits will allow sufficient UV transmission to cure an adhesive. This is an ideal situation since the process is fast, and no premature adhesive hardening can occur.

Once the components are attached, the circuit must be fixtured for soldering. A carrier board concept can be used, or circuits can be placed on a metal or high temperature plastic carrier or ''boat.'' The boat must be able to withstand the solder wave temperatures and present solder wetting. Plastics and non-solderable metals, such as titanium, are in use today. The boat must be designed to allow proper wetting of the assembly zone but limit solder contact where not needed. Proper circuit design, boat configuration, and process control even allow Mylar® circuits to be assembled. Polaroid camera circuits have been assembled with wave solder processing for several years. The amazing thing about the circuit is that only Mylar (polyester) base film is used, not polyimide. Figure 9.3 shows a Polaroid camera circuit, which has been wave soldered.

9.5.5 Special Soldering Techniques

Several approaches to minimize total circuit heating have been applied to flex in an attempt to allow the use of low cost polyester base. Polyester is generally adequate for most flexible circuit applications provided the rigors of soldering can be survived. One simple and very effective approach for simple assemblies is the hot bar method. The method can only be used for full wing devices or those where a heated bar can make contact with component leads. A heated bar or thermode is pressed against the upper surface of the component leads while the lower face of the leads are held against the circuit. Heat is transferred from the thermode to the lead, which either reflows solder on the lead or causes solder on the circuit to melt. Generally, solder plate on the circuit is sufficient to bond the component.

The hot bar process is extremely fast and minimizes total circuit heating very effectively. Polyester circuits can be used without fear of degradation. Copper-polyester calculator circuits have been assembled by this method since about mid-1970. In fact, the earliest SMT-Flex circuits were probably assembled by Texas Instruments in the U.S. The extraordinary fact is that polyimide was not used, but rather Mylar. This highly useful method, however, is very limited in scope. The component package must allow the thermode to contact the leads,

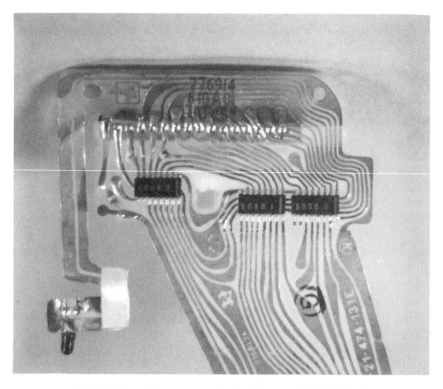

FIGURE 9.3   Polaroid camera circuit. (Courtesy of Sheldahl, Inc.)

and thermode contact is required for every device. This is still an ideal method for single component products, such as calculators.

The hot bar, or thermode method, can be used for fine pitch components, including molded TAB and conventional TAB. It is perhaps ironic that the first high volume product to use the concept on flex was the calculator circuit with 100-mil pitch components.

Solderless assembly methods have also been developed for flex to avoid the heat penalty of soldering. Common methods use Polymer Thick Film (PTF) technology in some form. Solder is replaced by conductive adhesive consisting of a conductive metal filler (occasionally carbon is used) in an organic polymer binder. The two basic types of conductive adhesives are isotropic, conducting equally in all directions, and anisotropic, where conductivity is direction dependent. Both systems are in commercial use today. Their application will be described here briefly. Chapter 11 covers the PTF field in detail.

Isotropic conductive adhesives are typified by silver-filled epoxies similar to the materials used for IC die attach. The conductive adhesive replaces solder paste, and thermal curing is used instead of reflow. The process is intrinsically clean and flux or CFC's are entirely eliminated. Here is how it works.

Silver conductive adhesive is stenciled onto circuit pads. The conductors can be either conventional etched copper or additive PTF printed ink. Components are then placed on the circuit with conventional pick and place equipment. Care must be used to place components more accurately and with the right amount of force. Conductive adhesive does not align a partially misaligned component liquid metal solder because surface tension is lower. Therefore, the adhesive printing and component placement must be accurate.

Once components are placed, heat will activate adhesive hardening or accelerate the process. Typical cure conditions range from 5–10 minutes and from 110–145°C. The hardened adhesive produces a low ohmic connection that is adequate for general use. Recent advances in this technology have produced highly stable junctions that are unaltered by heat and humidity. However, mechanical strength is lower than for metallurgical solder joints. Various mechanical enhancements, ranging from potting to shrink film overlays, can provide added strength for more stringent applications.

A more recent form of conductive adhesive is the anisotropic class. Conductors are distributed throughout a nonconductive binder at a concentration where individual particles are isolated. There is no electrical current pathway within the plane of the material. The adhesive, in liquid or film form, is applied to the circuit in the assembly area. Placement of the adhesive does not require any precision. The entire circuit can be covered. Components are placed with sufficient force to cause contact between the leads, the small conductive particles within the adhesive, and the pads on the circuit. The electrical path consists of tiny conductive particles trapped between the component lead and circuit pad. The small geometries of the conductive particles permit very fine pitch assembly. Conductor pitch, down to the finest yet created, can be connected with anisotropic adhesive. In fact, bare die have been connected in the laboratory by this method. The method will continue to gain popularity as finer pitch connections are required. The materials are especially suitable for use with flex because the circuit can conform to match the component and compensate for non-planarity.

9.5.6 Cleaning

Flex circuits can go through a wide range of cleaners without degradation, but adhesive degradation should be tested before purchase. Most flex manufacturers also have their own formulation of adhesive system so each system should be tested. While cleaning flex circuits, one will notice that the circuits float, or sprays and dryers blow the circuits off the belt. When the cleaning equipment is designed, make sure filters are placed before the pumps to eliminate pump damage if the flex circuit gets off the belt.

A strong move is on to eliminate CFC's as agreed in the Montreal Protocol.* Flex is perhaps easier to clean since the circuit under the component will flex slightly during spray cleaning. This allows better access under the component. Flex circuitry made by etching adhesive laminate adds the problem of adhesive surface and its tendency to trap solder balls and flux. However, newer adhesives, nonadhesive clads, and better solder mask placement are eliminating this problem.

9.6 DIRECT CHIP INTERCONNECT METHODS

Direct Chip Interconnect (DCI) concepts have been applied to flexible circuits. Level 2 interconnect processes, where there are two connections between the IC and the printed circuit, are in commercial use today. Level 1 interconnection, with only a single connection between the circuit and the IC, is still in the development stage. We will examine Level 2 interconnection next.

9.6.1 Wire Bonding

Wire bonding consists of forming an intermetallic bond between a thin gold or aluminum wire and the connection pad of an IC. The other end of the wire must be bonded to a lead frame or directly to a circuit. The Chip-On-Board (COB) concept requires that the IC be wire bonded to the circuit. This is a Level 2 interconnection scheme because there are two connections between the IC and circuit. COB wire bonding can be used with flex circuits, although it is not commonly done. One reason is that the flexible circuit, especially the common adhesive type laminate, absorbs much of the energy applied by the bonder. Rigid board does not really have an adhesive layer as does flex because the glass epoxy composite is self bonding. Flex, on the other hand, has a flexible adhesive layer, which can dissipate ultrasonic energy. A hot thermode of a thermocompression bonder can also cause delamination because of excessive heat. Wire bonding can be performed on conventional flex but often at a slower rate than rigid board.

More recent adhesiveless flexible circuitry is an ideal medium for high speed wire bonding. Lack of an adhesive allows higher temperature thermode settings and thus higher speed bonding. There is no adhesive to degrade, outgas, or otherwise degrade.

---

*An international treaty where the major manufacturing countries have agreed to completely phase out CFC's.

9.6.2 TAB Assembly

TAB components, where a miniature flex circuit becomes the "wire array" for mating the IC to the circuit, are ideal for assembly to a flex circuit. The TAB inner leads are connected to the IC pads and can be supplied to the assembler in reel form or as individualized components. The outer lead connection can be made in a conventional manner by solder bonding, using a hot bar. A newer method involves the use of anisotropic conductive adhesive. The adhesive can be applied to the circuit or to the TAB component. Heat and pressure complete the bonding process. This assembly method can be used on both conventional copper circuits and PTF types. The method on PTF circuits was pioneered in the U.S. and commercialized in Japan several years later.

Level 1 interconnection methods of flex have been studied thoroughly.[1] Most approaches use a "flipped" chip concept. The classical flip chip, or Controlled Collapse Chip Connection (C4) method is being adapted to flex. Flip chip IC's are constructed so that they can be directly soldered to circuits. The C4 type chip has precisely fabricated solder bumps on each pad. The metallurgy of the bump is fairly complex. The net result is that the bump softens, melts, and collapses in a controlled manner when heat is applied and, hence, the C4 nomenclature.

One limitation of the C4 process is that the micro dimensions of the interconnection create significant strain if the circuit's Thermal Coefficient of Expansion (TCE) is not closely matched to that of the chip. Typical circuits for flip chips are made of ceramic, which closely matches the expansion of silicon. Work with circuits made from polyimide has shown that the TCE for the common films, like Kapton, are too high for flip chip use. Although the compliant nature of the flex circuit can accommodate TCE mismatch between flex and an SMT package, the scale is too small for a flip chip. Because the bare IC die is an order of magnitude smaller than an SMD, the flex base film would need to be down-sized accordingly: this is not practical. A more viable solution is to use lower expansion polyimides that have recently become available. Copolymers made from selected polyimide monomers can be produced with TCE's as low as 4 ppm/°C. These materials can be used to produce circuits that are suitable for flip chips. Flex Flip Chip assemblies are now being tested for performance characteristics. See Chapter 8 for more on direct chip assembly.

Polymer Thick Film bonding materials and processes are also being applied to direct chip to flex. The predominant bonding material is fine pitch anisotropic conductive adhesive. Although materials can be easily made that will connect pads with only 2–4 mil spacings without shorting, the problem of a reliable connection still needs to be solved. Application of force to the top of the chip to create the connect appears to store stress in the adhesive film. The stress is relieved as the film expands in the Z-direction, and the connections made by the touching particles open up. Proper polymer rheology will solve this problem.[1] One additional technology for direct chip interconnect in Integrated Beam Lead. The concept, which is unique to flex, involves creating unsupported conductors within a flexible circuit. This is basically the creation of a TAB feature within a flex circuit. Because traditional flexible circuitry and 3-layer TAB (conductor/adhesive/dielectric base film) are essentially made with the same construction, fabrication of TAB features within the circuit is very practical. The most challenging obstacle is the creation of an opening or "window" within the circuit, placed so that cantilevered conductors are suspended over the opening. This construction allows a bare IC die to be placed inside the window with connection pads aligned with the cantilevered conductors or "fingers." The Inner Lead Bonding (ILB) process, involving a thermocompression bonding of leads to pads, is identical to classical TAB bonding. As with TAB, metal bumps (typically gold), must be formed on either the fingers or on the IC pads. Figure 9.4 shows TAB.

The advantages of Integrated Beam Lead over conventional TAB are several. Only a single bond between chip and circuit is used. This reduces the number of connections by 50%, which should reduce potential junction failure by the same amount. The required area is also smaller. On the down side, the ability to pretest is lost. One advantage of TAB is the easy ability to test the bonded chips before assembling the loaded TAB frame to a printed circuit. Since the

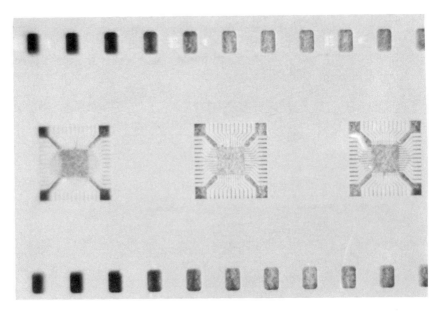

FIGURE 9.4   TAB

outer lead bonding step is eliminated, the IC is connected directly to the circuit with no intermediate stage where testing could be carried out. However, the ability to test IC's at wafer level has increased substantially, and many IC's can be pre-tested.

9.7 ASSEMBLY PERFORMANCE

9.7.1 Thermomechanical

As design engineers introduce new products and designs, thermal stress plays an important role in the selection of materials used. In today's applications, engineers are placing electronics in extremely active environments. Surface mount, like any other technology, must deal with the coefficient of thermal expansion. The expansion that once absorbed the leads on radial and axial leaded components is no longer available in surface mount. In SMT the solder joints are required to withstand this stress or have flexible base materials to absorb the load. Thermal cycling tests have been completed in which polyimide and polyester flexible circuits were compared to FR-4 board. The testing conditions were to Mil Spec., consisting of 1,000 thermal cycles from $-55°C$ to $125°C$ with a 30 second transition time. The test hardboards were constructed out of .031″ FR-4 material, and the flex circuits were .001″ Polyimide, with 1 ounce of ED copper. During the testing, some amazing discoveries were found. The hardboards saw failures in as little as 10 thermal cycles, and after 1,000 thermal cycles one or both terminals had visible cracks. The flex circuits at the same point had no failures. Testing was resumed on the flex circuit to 2,000 hours still with no failures, and only after 2,500 thermal cycles was a failure found. A second test was performed on the flex circuits using staking adhesives below the components. The staking adhesive degraded the life of the flex circuit in the thermal cycling test to 1,500 cycles before the first failure was found. Tests were also completed using selective rigidizing under the SMT components. The rigidizing also caused the flex circuits to fail after 1,500 cycles. Using both staking adhesives under the components and rigidizing, the flex circuits failed at 1,200 hours. Thermal cycling data led to the conclusion that the more rigid the base material, the sooner the solder joints would fail in a thermally active environment. The thinness, conformability, and compliancy of the flexible circuit makes it a truly unique product for component assembly. The compliant nature of the very thin base dielectric produces a nearly strain-free assembly, even when surface mount electronic components are used. This feature alone solves the solder joint thermomechanical fatigue problem, which plagues most rigid board constructions. This means that flex and SMT come together to produce an ideal synergy. Each technology supports and enhances the other. SMT-Flex assemblies can withstand thousands of thermal cycles. The strain-relief mechanism is superior to matched TCE concepts since localized heating does not cause problems for SMT-Flex.

Other benefits of the thinness of the dielectric base included excellent heat transfer and the ability to conform to housings and shaped backer boards. This means the flex assemblies can be used in 3-D configurations to maximize unit volume density. These important features of high junction reliability and assembly conformability have boosted the use of flex in a number of areas. We will explore basic design rules, assembly techniques, and actual applications in this chapter.

9.7.2 Quality

To many of us, flexible circuits are a new concept so we are unsure of the quality we can expect in the final package. Flexible circuits is one of the oldest circuit concepts ever documented, yet its overwhelming popularity is just beginning to surface. The auto industry has been using flex circuits since the late 1950's to improve quality in the dash assembly and reduce the use of interconnects. The calculator industry switched to SMT-Flex circuits over a decade ago to improve quality and reduce thickness. The camera industry has been using flexible circuits for a number of years, thus eliminating interconnect problems, and reducing the weight and physical size of the camera. The computer industry has been driving the read/write heads for years with flex circuits and within the past few years placed active components on the head assembly. Flexible circuits will improve the quality, reduce the weight, and reduce the physical size, if proper design techniques are used.

9.7.3 Cost Effectiveness

On a total assembly basis SMT-Flex is more cost competitive than hardboards. With the use of SMT-Flex, one will eliminate connectors, multiple boards, flat cable, and jumpers, along with the improved quality from elimination of interconnects. Added benefits will be reducing package size, weight, and solder joint fatigue. If all aspects are taken into account, the total package cost will be less for SMT-Flex than hardboards.

Figure 9.5 shows a flex circuit redesign in which eight connectors, three flat cables, and five hardboards were eliminated.

Figure 9.6 shows a circuit in which six connectors and four flat cables were eliminated. The final product was SMT-Flex with selective rigidizing and arms which tie to the outside world.

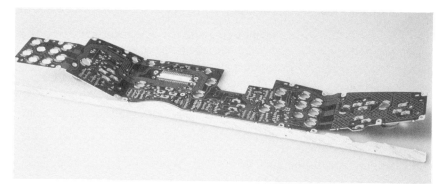

FIGURE 9.5   Integrated assembly A. (Courtesy of Sheldahl, Inc.)

FIGURE 9.6   Integrated assembly B. (Courtesy of Sheldahl, Inc.)

9.8 CONNECTORS

Flex often serves as a dynamic interconnect cable between two or more points. This long history as a cabling material has led to the development of numerous connectors specifically for flex.

9.8.1 Insulation Displacement (Piercing) Types

The thin base film used to manufacture flex allows connector designs to be used, which pierce through the dielectric. The insulation displacement type can be further divided into those that form a mechanical contact against the conductor and those that actually pierce the conductor. Connectors used for polymer thick film usually penetrate the conductor to form a gas tight seal. Both individual terminals, called crimp on type, and molded plastic array connectors are widely available.

9.8.2 Solder Type Connectors

There are two basic types of solderable connectors; feed-through and surface mount. Feed-through connectors require that the circuit be designed with holes in the dielectric. The connector is then applied from the back side of the circuit and hand or wave soldered to the conductor side. Polyimide is commonly used, but polyester will usually suffice because heating is fairly localized. When polyester is used, the soldering zone must be carefully localized by use of a wave solder carrier to ensure that excessive base film shrinkage does not occur.

Surface Mount connectors can be applied like other SMD's. The common reflow heating methods, such as IR, vapor phase, and hot bar work well for connectors.

9.8.3 Zero Insertion Force Connectors

A number of clever Zero Insertion Force (ZIF) connectors have been designed for use with flex. Basic concepts include a mechanism for clamping the flex conductors to either the connector or against a rigid board. Force can be generated by a spring mechanism or a force generating lever. Most membrane keyboards use a ZIF connector scheme.

9.8.4 Flex as a Connector

A flex circuit can be designed to mate with another circuit or a device without requiring the addition of a connector termination. The end intended for connection is usually plated with solder or gold. PTF circuits, made with silver ink, do not need additional protection although carbon ink may be placed over the silver to preclude the chance of silver migration.

9.9 TEST AND REPAIR

9.9.1 Testing

Flexible circuitry is slightly more difficult to handle for testing than rigid board, especially if tested in a multicircuit array. Pre-blanked circuits tend to move in their positions often requiring sets of alignment pins to position "sets" of circuits. Vacuum hold down is very effective for flex since the circuit will easily conform to a test probe array.

9.9.2 Repair

Polyimide circuits can be repeatedly resoldered without concern for degradation. Many laminates conform to the "5 times resolder test" criterion. The same techniques used for hardboard repair can be applied to flex. It may be necessary to use vacuum hold down when large IC's are removed.

Polyester circuits are not generally considered to be repairable, although exceptions exist. Application of heat to a polyester circuit can cause softening and shrinkage of the base film. Remelting techniques that restrict heating to the component zone are best.

Polymer Thick Film circuits may be repairable, depending on the adhesive used to bond components. Some adhesives can be brought to a softening point without damaging the substrate, which is typically polyester. Direct contact heating, such as "thermal tweezers," give best results. Unless there is reason to believe otherwise, PTF circuits should be assumed not repairable.

9.10 DESIGN CONSIDERATIONS FOR ASSEMBLY

The SMD land patterns established for hard board by the IPC[3] generally apply to flexible circuit assembly. Land patterns may vary somewhat for flex, and this can be product specific. However, there are some subtle and also a few major differences that apply specifically to flex. Conductor routing will require greater detail on flex to eliminate conductor cracking in the flexing zone. A common error is to enter a pad with a conductor at the wrong location. Figure 9.7 shows the correct design.

Figures 9.8 and 9.9 show how common errors can cause the component to skew during manufacturing.

Solder leaching is another commonly overlooked problem, which can cause numerous man-hours of rework. The cover layer or cover coat should dam the conductor to eliminate the solder from leaching. Some assemblers modify the stencils to place solder paste on the exposed conductor, which will eliminate the solder leaching, thus, an adequate amount of solder is present to form a good solder joint.

Fiducial location and type of mark are also important for machine accuracy. Common types of fiducial marks are circles, squares, rectangles, butterflies, diamonds, etc. Figure 9.10 shows common examples of fiducial marks.

Each machine on the market today has its limitations on fiducial mark sizes,

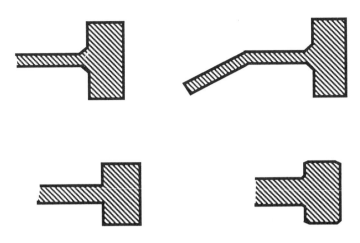

Acceptable conductor routing

FIGURE 9.7   SMT pad and conductor.

shapes, and processing ability. Fiducial mark sizes and spacing surrounding the marks are requirements of the equipment manufacturer. Fiducial marks are not standardized in the industry, thus a fiducial mark that can be read by one piece of SMT equipment might not work on the next piece of SMT equipment. Vision systems that read the fiducial marks are also sensitive to plating types; i.e., gold, plated solder, copper, and polymer thick film. A fiducial mark that reads perfectly good on copper might be impossible to read as a gold mark. *Automated equipment is not tolerant of contaminated fiducial marks.* Finger prints, oxides, stains, adhesive smears, cover coat smears, all can cause equipment shutdowns if the SMT equipment cannot handle the change in fiducial reflectivity. The most difficult fiducial reading problem on flexible circuits is the planarity of the circuit. Unlike FR-4 boards, flex circuits do not lay flat. The planarity differences from fiducial to fiducial are enough to shut-down most manufacturers equipment on the market today.

Electrical test pads and locations also need consideration during the design stage, if electrical testing is to be performed on the final assembly. Electrical tester pad sizes are critical, if low testing failure rates are to be achieved. The following performance comparison for hardboard test pad sizes may be used as a guide.

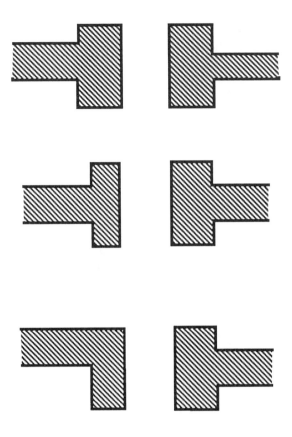

Incorrect conductor-pad arrangements

FIGURE 9.8   Incorrect conductor routing.

TEST PAD SIZE (DIAMETER)	TARGETS MISSED HP SIMPLATE FIXTURE	CONVENTIONAL FIXTURE, PPM
.040″ DIA	.02 PPM	1.20 PPM
.035″ DIA	1.30 PPM	29.99 PPM
.030″ DIA	48.00 PPM	465.00 PPM
.025″ DIA	1002.00 PPM	4849.00 PPM

Many designs utilize the configurable properties of flex to create interconnect "cables." The assembly area should be as localized as possible so that bonding will not occur under components. It may be useful to consider a selectively

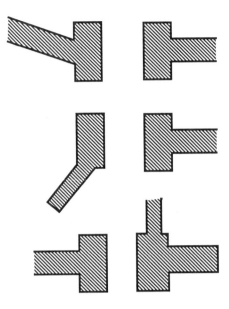

Unsymmetrical arrangements cause skew

FIGURE 9.9   Incorrect conductor routing example.

rigidized region for components. When the population zone is rigidized with a stiffener, no carrier board is required during the assembly processes. Both non-metal and metallic stiffeners can be used depending on the desired performance. Use of aluminum backers has become popular because shielding and excellent thermal management will result. Thermally conductive backer adhesives are available for particularly challenging thermal management situations. Thinner base dielectric or polyimide filled with thermal conductors can be used for demanding situations. Heat dissipation factors superior to all but high thermal ceramic boards can be achieved. This high heat transfer construction strategy is especially attractive for automotive and power control circuitry.[2] Table 9.2 shows some heat transfer values for various backer and construction materials.

9.11 APPLICATIONS

SMT-Flex can be used for virtually any application, after the requirements are defined. The requirements will define the operating environment and temperatures the assembly is expected to see. A product that is going to remain in the office and not see thermal swings should consider Polymer Thick Film Tech-

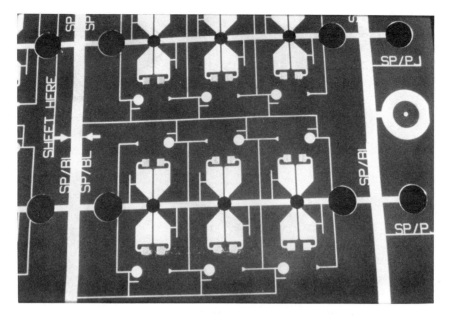

FIGURE 9.10   Common fiducial marks. (Courtesy of Sheldahl, Inc.)

TABLE 9.2   Thermal Transfer for Electronic Materials

Material	Thermal Conductivity (W/m·K)
Aluminum	236
Kapton	0.15
Acrylic Adhesive	0.20
Copper	398
Alumina	20
Beryllium Oxide	275

nology, in which the conductors are conductive ink, and the components are mechanically bonded with conductive adhesive. Products that need higher performance electrically and mechanically should consider polyimide base, electrodeposited copper, and eutetic solder. Products that require millions of flexing motions require rolled annealed copper, and products that are designed for resistance soldering, or the highest end requirements, should consider adhesiveless products.

Figure 9.11 shows an automotive air conditioning sensor for flex to install, under the hood application, while Figure 9.12 shows the fully assembled product in a cylindrical housing. Note the excellent use of conformability. The construction is polyimide base with ED copper and UV cover coat.

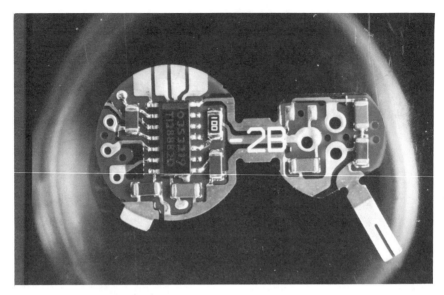

FIGURE 9.11   Air conditioning sensor. (Courtesy of Sheldahl, Inc.)

FIGURE 9.12   Final assembly-pressure sensor. (Courtesy of Sheldahl, Inc.)

The use of SMT-Flex in the automobile industry has seen outstanding growth due to the reduction of thermal fatigue on solder joints. Figure 9.13 shows an automotive ignition sensor, and Figure 9.14 shows the assembled module. The application is flex to install, in a extremely active thermal environment. The material selected is polyimide base, ED copper, and polyimide cover layer.

The communication industry is beginning to use SMT-Flex to reduce package size and allow backlighting. Figure 9.15 shows a hand-held cellular telephone flexible circuit, and Figure 9.16 shows the finished assembly.

Due to the backlighting requirements, polyester was selected. The thermal degradation temperature of polyester is 125°C, and the reflow temperature of eutetic solder is 181°C: thus, requiring dedicated processing equipment. The material selection is polyester base, ED copper, and Polyester cover layer. The circuit also has gold plated keypads and solder plated connector areas.

The computer industry has been using flex circuits to improve I/O times on disk drives by placing the active electronics on the read/write heads to allow faster access times and shorter response times. Figure 9.17 shows a computer disk drive circuit, which has been designed for millions of flex cycles along with carrying active SMT components. The material selection is polyimide base, rolled annealed copper, and polyimide cover layer.

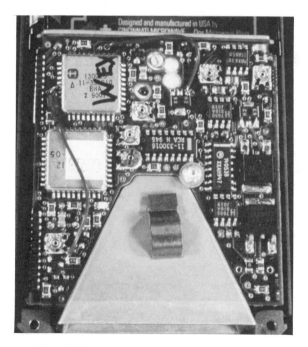

FIGURE 9.13   Ignition sensor. (Courtesy of Sheldahl, Inc.)

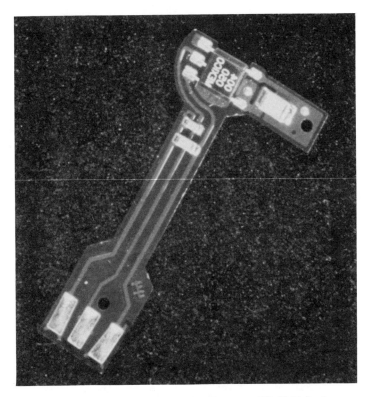

FIGURE 9.14   Finished assembly. (Courtesy of Sheldahl, Inc.)

The computer industry also uses Polymer Thick Film for keyboard applications. Additive processing allows the manufacturer to produce a more cost effective circuit, while eliminating the environmental hazards of etching and plating. Figure 9.18 shows a Polymer Thick Film circuit designed for keyboard applications complete with encoding electronics. The material selection is polyester base, screened conductive ink conductors, and conductive adhesive for component assembly.

Some designs require component zones in more than one location. Some disk drive and printer cables are most efficiently designed with decoding and encoding at each end of the circuit. Mixed technology, especially SMT and TAB or direct chip, are often desirable for disk drives and other applications where a very light component array is required at the dynamic end of the circuit. Very light stiffeners can be used for dynamic ends. These include backers made with polyimide base films as well as thin molded plastics. Some designs bond a

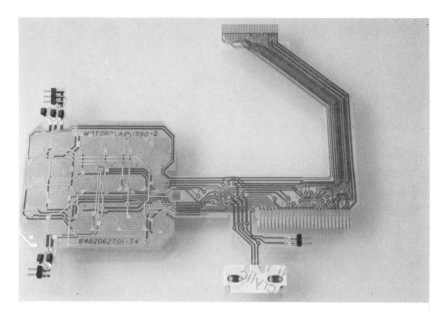

FIGURE 9.15   Cellular telephone flexible circuit. (Courtesy of Sheldahl, Inc.)

mounting fixture to the circuit, which serves as both a stiffener and a mounting/ alignment device.

Designs that require both SMD's and feed-through components should be laid out so that each component technology is isolated. This permits wave and reflow soldering to be accomplished on the same circuit quite easily. An ideal strategy involves first completing the SMT assembly. The SMT section can be bent out of the way to allow the feed-through components to be wave soldered. A simple boat or wave solder carrier is used to ensure that the SMT area is protected and that no solder remelting occurs. Figure 9.19 shows an early mixed design that was used for the first ''smart relay.'' This design did not use stiffeners but additional knowledge gained since the 1981 implementation would strongly recommend stiffeners, especially for the heavy feed-through relays.

Another consideration is the array strategy for small circuits. Because typical flexible circuit image sizes can meet or exceed $18'' \times 18''$ areas, a multi-up pattern must be used in producing the individual circuits. Although individual circuits can be blanked out for assembly, a more efficient strategy can be to assemble the full or partial array. An $18'' \times 18''$ area can contain hundreds of circuits and maintaining the full sheet as long as possible can increase efficiency substantially. The entire process including testing can be accomplished in the full sheet form.

Use of full sheet arrays for assembly requires that individual circuits be either

FIGURE 9.16   Finished assembly. (Courtesy of Sheldahl, Inc.)

blanked after assembly and test or pre-blanked by the circuit manufacturer. There are at least three possible concepts in use today. Circuits can be partially blanked out with small holding tabs left in place. This blanking concept is referred to as notched die and also "perf" (for perforated) die cutting. After assembly, the circuits can be plucked out of the sheets. Various circuit manufacturers have proprietary tooling technologies for producing the holding tabs. Another concept involves blanking the circuits on a carrier sheet. A low tack adhesive holds the circuits in place until assembly is complete. A third method

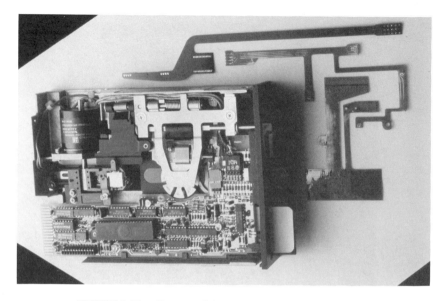

FIGURE 9.17   Computer disk drive. (Courtesy of Sheldahl, Inc.)

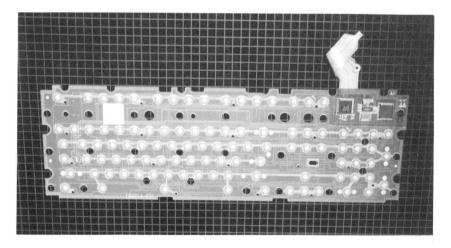

FIGURE 9.18   PTF encoded keyboard. (Courtesy of Poly-Flex Circuits, Inc.)

involves the use of selective stiffeners. Although the method, pioneered by Sheldahl and Ford Motor Company, was designed for feed-through assembly, it is very applicable to SMT. The basic concept is to selectively print backer adhesive onto a precut backer. The circuit is then bonded only where the ad-

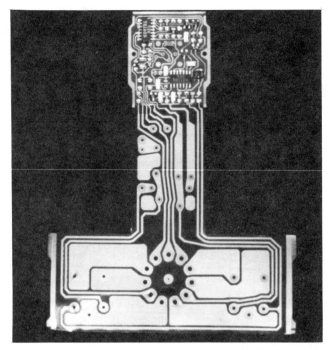

FIGURE 9.19   Smart relay. (Courtesy of Sheldahl, Inc.)

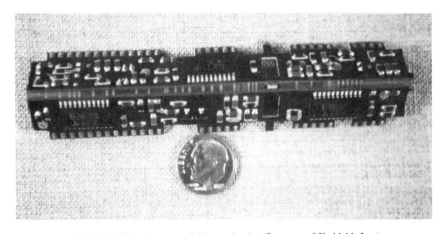

FIGURE 9.20   Japanese foldover circuit. (Courtesy of Sheldahl, Inc.)

hesive has been applied. The backer cuts are designed so that individual circuits can be snapped out after assembly, similar to ceramic hybrid "snap strate" concepts. Some designs have even been implemented with aluminum backers as seen in Figure 9.20.

SMT assembly, wave soldering in this example, the circuits are cut out of the array. Note that the circuit is folded over to produce a pseudo double-side edge card.

The fold-over designs, unique to flexible circuits, are an excellent way of reducing apparent circuit area by 50% or more. The backer can be designed to create one or more flex circuit hinges.

SUMMARY

Flexible circuitry is a unit electronic assembly medium. SMT and flex are mutually synergistic and the combination provides a very reliable high density assembly. The 3D characteristics of these assemblies allow a very efficient use of volume and a high level of integration. Direct chip interconnect schemes are at a unique advantage with flex because of the compliancy of the thin circuit, the low thermal expansion properties of new polyimides, and the ability to fabricate TAB features within the circuit.

References

1. Gilleo, K. Direct Chip Interconnect Using Polymer Thick Film Technology, Electronic Connection Conference, IEEE, Houston, TX, May 1989.
2. Gilleo, K. Expanding Power Hybrid Capability with Flexible Circuitry, *Electronic Manufacturing*, Sept. 1990.
3. Surface Mount Land Patterns ANSI/IPC-SM-782.

10

Polymer Thick Film Flex

Chon Wong

Poly-Flex Circuits, Inc.

10.1 INTRODUCTION

The Polymer Thick Film (PTF) circuit assembly is typically found in such applications as alpha numeric (A/N) keyboards, telephone circuits, calculators, medical products, printers, and other consumer electronic products. Examples of the PTF processes, electrical and material characteristics are presented to give the flavor of this interesting, modern technology. PTF technology uses an additive concept compared to the traditional subtractive copper printed circuit methodology. The conductor patterns are screen printed onto the substrate instead of chemically etching away metal laminate. Details of this additive process will be presented in the next section. The PTF circuit in its simplest form can be just a conductive pattern on an insulating substrate, which serves as an electrical bus carrying current from one point to another.

A popular PTF product is the membrane switch, a simple array of conductors on a flexible insulating substrate. A single membrane circuit is folded over a perforated spacer sheet, which keeps switch contact elements separated until force is applied to close the switch. The spacer can be a thin sheet of dielectric with many holes (about 0.5 inch in diameter) looking like a slice of Swiss cheese. The hole diameter and the spacer thickness determine the force necessary for switch closure. There are many design variations of the membrane switch, but the two most common designs are the "foldover," just described, and the shorting pad type. The shorting pad design keeps all of the circuitry on a single layer. The switch element consists of a "comb" of interdigited "fingers" where a separate short out contacter closes the lower half switch. This shortout design can make use of silicone rubber pads with electrically conductive disks. Although still classified as a membrane switch it really is not, because there is no thin spacer or membrane separator in the design. Figure 10.1 shows a membrane switch commonly used in computer keyboards.

The foldover design uses large conductive dots or round pads (usually 0.25 inches in diameter) opposing one another through the hole in the spacer. Since PTF materials are flexible, a single circuit can be printed and folded over to sandwich the spacer. Maximum design reliability is obtained by folding the conductive traces over a minimum radius so that a high stress sharp bend does not occur. The UL796 method for measuring the performance of the fold is done by monitoring the increase in resistance of the trace over a 0.25 inch diameter mandrel as it is flexed 100 times. The short out design, described earlier, can also be constructed of hardboard while the foldover must be made with a flexible circuit.

The membrane and other PTF switches have been commercial for over 15 years and constitute the largest part of the emerging PTF flexible circuit market. The foldover membrane switch requires a highly durable and flexible circuit by the nature of its design. The conductive ink traces must be able to bend over a fairly tight radius without cracking or losing conductivity. Modern PTF conductive inks are highly flexible and can be bent into a tight fold, although a moderate radius of .060″ or more is recommended to reduce stress.

Some may not wish to classify the membrane switch as a true circuit because most designs have not utilized electronic components. However, most copper flexible circuits still do not have active components, serving only as a flexible

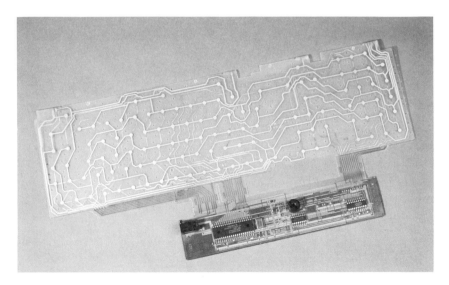

FIGURE 10.1   Membrane switch. (Courtesy of Poly-Flex Circuits, Inc.)

cable or jumper. Adding components to switches has been an increasingly popular trend as designs strive to integrate features. Starting with the addition of simple LED's, the switch has grown more sophisticated to a level where encoding electronics is now added to the keyboard directly. Figure 10.2 shows a telephone assembly with components bonded directly to the PTF switch/circuit. Note that a flexible cable was designed into the flexible circuit, thus eliminating connectors and a separate cable.

The advanced PTF circuits includes the mounting of electronic components on circuit assemblies. The interconnect between the conductive traces and components can be achieved with conductive epoxies or anisotropic conductive adhesives. Figure 10.3 shows a A/N fully encoded keyboard where electronic components have been bonded using conductive adhesive. The launching of SMT in the PCB industry has reduced the size of electronic components and allowed denser PCB's. It has also helped the PTF industry because these SMD's do not have axial leads that require feed-through holes. The SMD is an ideal package for use with conductive adhesive because the component can be applied

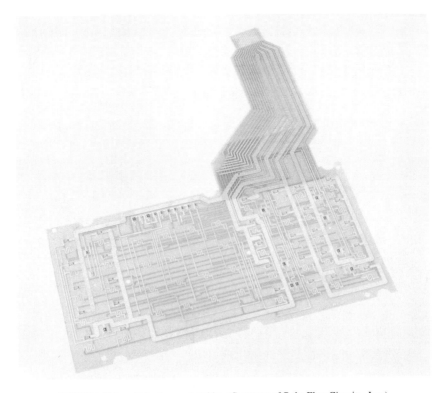

FIGURE 10.2   Telephone assembly. (Courtesy of Poly-Flex Circuits, Inc.)

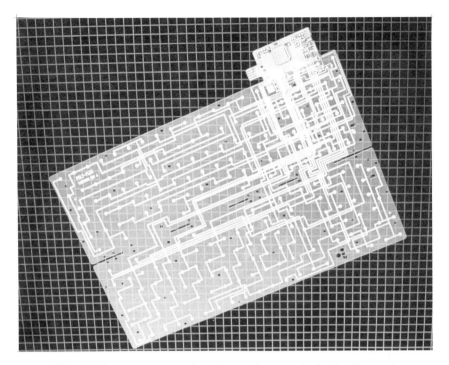

FIGURE 10.3   PTF encoded A/N keyboard. (Courtesy of Poly-Flex Circuits, Inc.)

to circuit pads, which have been stenciled with conductive adhesive. A butt, or lap joint, is formed with good electrical and mechanical performance. An adhesive called Poly-Solder® was introduced in 1989, resulting in the commercialization of PTF assemblies. The material provides stable junctions with standard SMD's thus making polymer bonding a viable alternative to soldering.

Higher density is readily achieved in more advanced PTF circuits. A dielectric ink is incorporated to give the circuit a multilayer structure. Vias are created in the circuit by leaving openings in the printed dielectric, which are placed over traces on the lower conductor layers at the desired connection locations. The via hole within the dielectric becomes filled with conductive ink during the next printing step. This creates electrical connections between conductors on different layers. The process of printing conductors, dielectric and conductor, can be repeated until the desired number of circuit layers has been achieved. Although four conductive layers is today's practical limit for volume production, up to a dozen layers have been built for test prototype circuits.

An alternate technology is the Printed Through Hole technique where the holes are fabricated in the substrate before printing. The conductive ink is printed through the hole creating a PTF "barrel," or a printed through hole. The end

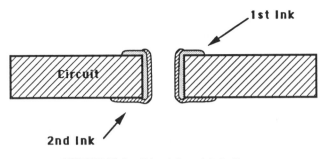

FIGURE 10.4   Printed through hole diagram

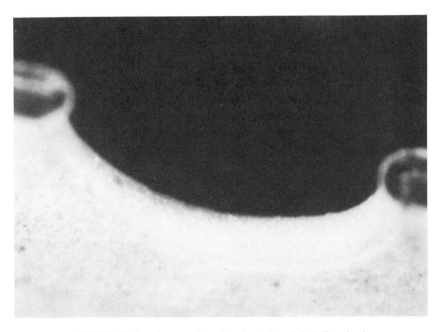

FIGURE 10.5   Cross-section of a printed through hole flex circuit

result looks just like a copper Plated-Through Hole. The Printed Through Hole connection can be used with other circuit technologies, such as etched copper. Conductive PTF ink can be printed on the back side of a single-sided copper flex circuit. Use of controlled vacuum[1] allows the ink to be pulled onto the copper, forming an annular ring junction. This process is used in high volume to produce calculator circuits. This hybrid allows the designer to take advantage of the features of copper, like solderability and high current capacity, while enjoying the simplification and cost benefits of PTF.[2] Figure 10.4 shows a diagram of a Printed Through Hole while Figure 10.5 shows a cross-section of the interconnect in a flexible circuit.

10.2 THE POLYMER THICK FILM ADDITIVE CIRCUIT PROCESS

The PTF additive process described here is a general process. Conductive particles, such as silver, copper, or carbon, are dispersed in a polymeric binder. The binder can be a solvent-borne resin or a thermoset. Silver inks continue to enjoy a high level of popularity because of the excellent balance of properties they provide. The most important property of silver is that it remains highly conductive even in an oxidizing environment. Copper forms an insulative oxide layer and must be compounded with antioxidants to have any chance of remaining conductive after heat aging.

These inks are applied to flexible dielectric films such as Polyester, Polycarbonate, Polyimide, and FR-4 epoxy. Although FR-4 board is generally considered a rigid board material, newer thin versions, down to 4 mils, are flexible enough to run in a continuous roll operation. Thin FR-4 is commercially available in the U.S. and Japan, and it is likely that some flex plants are processing the material in roll-to-roll format.

These conductive inks are almost always applied to the base film substrate by screen printing. Temperature stabilized polyester film is commonly used because these films have very small lateral distortion, nominally below 0.10 mil per linear inch. Unstabilized polyester will usually have too much shrinkage to be used for multiple printing operations. The thickness of the polyester film substrate comes in standard thicknesses of 3, 5, and 7 mils, although virtually any thickness can be obtained from the dozen or more world suppliers. Thickness choice is governed by the requirements of the overall mechanical assembly, stretch elastic characteristics, and its flexibility. The base film thickness can be critical in switches and also in assemblies because the film forms an integral part of the PTF circuit assembly.

The dielectric ink is the next critical ingredient in fabricating a complete PTF circuit. The dielectric has several possible functions. One purpose is to serve as a "solder mask" which helps contain the printed or stenciled adhesive while also protecting the terminal conductor traces or those passing between pads from excess adhesive. A second function is to protect the entire circuit from the environment and to isolate conductors from external contact. A third use is to provide electrical isolation when one conductive trace crosses another conductive track. It is a common practice to use two screen passes of dielectric when it is used to isolate layers. This greatly minimizes the chance of a pinhole defect creating a short between the two conductive layers. A typical dielectric ink thickness is from 12–25 microns (.0005 to .001″) per pass so that the total thickness may be up to 50 microns (.002″).

The dielectric ink must have very good barrier properties to prevent metal electromigration from occurring, when power is applied under harsh environmental conditions. Since the typical PTF conductive ink is made from silver, a phenomenon known as silver migration can cause electrical shorts between conductors when exposed to liquid water and a DC voltage. The electromigration is essentially an electroplating mechanism where silver ions dissolve in water and migrate to the negative pole (cathode) of the circuit. The silver ion is reduced back to silver metal when it picks up an electron from the cathode. The tree-like dendrite of plated out silver can eventually bridge the gap between the two adjacent conductors. Our testing has shown that the problem of migration can be eliminated by preventing liquid water from contacting the silver or by applying a dielectric that is too dense to permit the silver ions from migrating to the opposite electrode.

A common antimigration design involves covering exposed silver conductors with dielectric. Obviously, all silver conductors cannot be covered. Switch pads and other contact points can be connected to non-silver pads or overprinted with carbon ink. Since carbon ink is several hundred times more resistive than silver ink, caution has to be taken in using pure carbon pads without silver backing since the resistance of the switch interdigited pads may exceed the electrical resistance requirements of the matrix detection scheme. The SMD mounting pads are usually made with silver ink that is left exposed. Most IC's are designed so that power and ground are at opposite ends of the package, and this is enough separation to prevent migration.

The electronic components used to decode a switch matrix often uses CMOS (Complementary Metal Oxide Semiconductor) technology. They are highly sensitive to high voltage especially from electrostatic charges. Electrostatic voltages may come as a surprise to many people for its units of measure is in kilovolts (a kilovolt is a thousand volts). On a dry day, combing your hair can give rise to several kilovolts. Electrostatic damage on electronic components is a critical issue to the electronic industry. One solution is to cover the electronic components with conformal coating. The membrane switch sections are automatically protected from Electrostatic Discharge (ESD) because the conductive tracks are protected by the polyester substrate. Sometimes 5- or 7-mil thick polyester is chosen on this basis.

After understanding the additive process, it is important to look at the PTF materials characteristics in particular.

10.3 PTF MATERIALS CHARACTERISTICS

10.3.1 The Silver Loaded Polymer Ink

As mentioned earlier, conductive inks are generally made by loading silver into a polymer ink binder. The most common type of silver ink uses a thermoplastic

TABLE 10.1   PTF Materials

A) PTF Conductive Ink
    1) Thermoplastic
    2) Thermoset

B) PTF Dielectric Ink
    1) Solvent-thermoplastic
    2) Thermoset
    3) Radiation curable

C) PTF Resistors

D) Bonding Agent
    1) Isotropic,
    2) Anisotropic,
    3) Nonconductive

E) Conformal Coatings/Bond Enhancer

F) Base Dielectric

binder in a solvent. This solvent allows the highly silver loaded ink to be screen printed with the right viscosity and sufficient shelf life. Solvent type inks cure rapidly since only the removal of solvent is required. The silver particle distribution must stay within a certain particle size range for consistent properties and yet be small enough to pass through the finest printing screen mesh used.

In the process of curing the thermoplastic ink, the solvent is driven out, leaving the polymer resin matrix and the silver particles. Optimum conductivity of the silver trace requires that virtually all of the solvent be driven off. A properly cured silver ink has a fairly good electrical conductivity but not as good as etched copper. Although silver is a slightly better conductor than copper, current must pass through thousands of silver particles and their mechanical junctions. This adds resistance so that the best silver ink is about 10 times more resistive than continuous copper.

The adhesion of the conductive ink to either the substrate or the dielectric material is measured by the ASTM method D3359-83 or the IPC-TM650. In this method an adhesive tape is used to adhere on the grided ink surface and attempts to remove the ink by lifting ink off. In addition an abrasion test can be applied using a rubber eraser. UL796 test method uses a medium grade (grade 1/2) abrasive cloth to erase for 60 cycles.

In addition to the conductive ink bonding, the electrical property should be consistent. The UL796 test is designed with 0.5 ampere flow through the circuit substrate strip under a tension force of 0.5 pounds. The test strip is also subjected to flexing over a 0.25-inch radius fixture by bending back and forth at an angle of 180° for 50 cycles or until the circuit is open or the substrate cracks, splits, or delaminates. The IPC-TM650 method performs a similar flexural fatigue and ductility test.

Electrical volume conductivity is an electrical parameter that describes the ease of current flow in a material with units of Siemens per meter (S/m). Resistivity, the reciprocal of conductivity, is more commonly used. Units are ohms-meter (ohm.m) and can be determined by using a four point probe micrometer on a known dimension strip of conductive ink. A stencil can be used to reproduce a conductive strip of definite dimensions. Silver and copper have conductivity values of 6.10×10^7 S/m and 5.80×10^7 S/m, respectively. However, in the PCB industry the term ounce per square foot of copper is being used instead of thickness. An ounce of copper per square foot is equivalent to a laminate thickness of about 1.4 mil. This equivalent way of expressing copper laminate thickness can be understood from the density of copper. A sheet of copper laminate with length and width of a square foot weighing one ounce will have a thickness of approximately 1.4 mils or 35 microns.

Resistance of copper conductors on sheets of copper laminate with uniform thickness can be estimated easily. Sheet resistance can be calculated geometrically from the copper conductivity relationship. See Eq. 10.1.

$$\sigma_v = \frac{\text{length of track } (m)}{\text{resistance } (\Omega) \times \text{thickness } (m) \times \text{width } (m)} = \text{siemens/meter} \quad (10\text{-}1)$$

Another common parameter is sheet resistivity. This parameter enables one to estimate the conductor resistance easily by counting the ratio of number of squares along its length and width and multiply it by R_{square} to give resistance. The sheet resistivity of copper laminate is given by Eq. 10.2

$$R_\square = \frac{1}{\sigma t} = \frac{1}{5.8 \times 10^7 \times 1.4 \times 2.54 \times 10^{-5}}$$

$$= 0.00048 \ \Omega \text{ per square (assuming 1.4 mil thick)} \quad (10\text{-}2)$$

If this parameter is normalized to 1.0 mil, the sheet resistivity would be 0.00067 ohms/square (1 mil) or 0.67 mohm/square/mil. This is about 22 times more conductive than today's common commercial silver loaded inks specified at 15 mohms/square/mil. However, some of the experimental inks are capable of achieving under 7 mohm/square/mil. The trend toward more conductive silver inks will continue.

Another important material characteristic is the Glass Transition Temperature, T_g. This temperature can be interpreted as the phase transition from glassy to plastic temperature of polymer binder in the silver ink. This temperature can be determined by using a Differential Scanning Calorimeter (DSC) instrument where temperature is monitored on both a loaded fully cured sample and a blank sample. The silver ink loses its rigidity in mechanical form once exceeding this temperature. In other words it behaves more like a liquid than a solid exceeding this temperature. The silver ink will be deformed easily with external forces at this temperature. The IPC-TM650 also discusses the measurement technique for T_g and Z-axis expansion.

10.3.2 Conductive Carbon PTF Ink

The use of carbon particles in place of silver particles has two important consequences. Resistance is increased by three orders of magnitude. Although the resistivity of carbon ink is high it can be varied from 40 ohms/square to 100K ohms/square. This wide range of resistivity allows good carbon resistors to be printed. Many of the commercially available carbon inks can be blended with various percentage weights of silver to give different resistivity. A second important difference for carbon inks is inertness. Unlike silver, carbon ink does not undergo electromigration. This makes carbon ink a good choice for switch contact pads where there is a potential for metal migration. The carbon ink may be printed over silver ink.

One important electrical parameter of resistor is the Temperature Coefficient of Resistance (TCR), which measures the increase in resistance due to temperature rise. A positive value of TCR indicates an increase in resistivity when temperature rises, whereas a negative value of TCR indicates a decrease. The definition for TCR is given by Eq. 10.3

$$TCR = \frac{(R(T_2) - R(T_1)) \times 10^6}{R(T_1) [T_2 - T_1]} \text{ ppm/}^\circ\text{C} \qquad (10\text{-}3)$$

where

$R(T_2)$ = the resistance at the higher temperature T_2
$R(T_1)$ = the resistance at the lower temperature T_1

The electrical conductivity of carbon-polymer composites does not increase linearly with volume amount of carbon particulate present. There is little or no conductivity when the percentage volume of carbon black falls below 20% in general. Conductivity increases rapidly displaying the nonlinear conduction behavior as the percentage volume increases. There are different types of carbon blacks available, as well as graphite powder. The DC conductivity of carbon ink usually increases as temperature increases (positive TCR).

10.3.3 Substrate and Dielectric Ink

The important properties required by a substrate are common to the dielectric film with the exception that it must be capable of being uniform and flexible while being a thin film:

a   Tensile strength and tensile modulus in all axes
b   Flexibility or rigidity
c   Abrasion resistance and adhesion
d   Low and uniform thermal coefficient of expansion
e   Insulating and dielectric property
f   Flammability
g   Chemical inertness

The substrates that can be used are Polyethylene Terephthalate (PET), Polycarbonate, Polyetherimide, Polyimide, and various high performance films. The tensile strength determines the ultimate failure of the material under stress whereas the tensile modulus determines the material's behavior under stress, including flexibility. Many of the PET films used in membrane switches are planar oriented films where in their manufacturing process involves stretching in the plane of the film. Therefore their tensile strength in the normal direction to the film plane is much weaker. This can reduce bond strength of attached components.

The folding endurance of the substrate is related to its flexibility. An example of the folding endurance test was given in Section 10.2.1. It measures the number of flexing cycles at a given actuating force that a substrate can withstand before mechanical and electrical failure occur. The abrasion resistance and adhesion quality determines how much abrasion force is required to remove conductive tracks adhesion on the substrate and also how well the substrate can withstand abrasion mechanically. The T_g is related to its thermal coefficient of expansion. A high T_g material is usually one that has a low thermal coefficient of expansion. For dimensional stability, polyester should be heat stabilized before printing. The stabilization temperature must exceed the processing temperature by more than 15 to 30°C in order to remain relatively free of dimensional changes during processing. Normally this temperature is close to T_g. Polycarbonate is normally processed under 130°C, and it does not require heat stabilization because it has good dimensional stability below 130°C. Polyimide and polyethersulfone have extremely high stability because their T_g value is exceptionally high. Processing below 500°F does not present any problem. These materials have extremely good chemical stability and are also inert.

Dielectric insulating properties of both the substrate and circuit interlayers are important because they effect the insulation properties, such as surface insulation resistance, dielectric breakdown voltage, permittivity, and dielectric loss. The ASTM method D150-81 describes the AC loss characteristics and permittivity of these materials. DC insulation resistance can be performed using the ASTM 257 method. The relative dielectric constant for PET films is about 3.5, whereas polycarbonate is 2.9. The dielectric strength measures the dielectric breakdown properties of the material, and it is expressed in volts/mil.

The flammability of the substrate is critical in determining the ease of ignition, burning rate, flame rate, intensity of burning, and products of combustion. One criteria of the flammability test can be found in the UL 94 test method. The flammability tests are separated into horizontal and vertical categories. The horizontal burning test classifies materials for 94HB, 94HBF, 94HF-1, or 94HF-2 whereas the vertical burning test classifies materials for 94-5VA or 94-5VB, 94VTM-0, 94VTM-1, or 94VT M-2. The VTM tests determine the flame spread index. The HF tests are for foam materials.

10.3.4 Bonding agent

In a PTF circuit, the components are bonded to the substrate and conductors. There are principally two kinds of bonding agents. One is isotropic and the other anisotropic. The isotropic agent is one where the electrical conduction path of the bond is independent of direction, and conductivity is uniform in all directions. However, there are bonding agents that are anisotropic, or Z-axis conductive adhesives, and the electrical conduction path is unidirectional. These Z-axis bonding agents are made of highly dispersed conductive particles in a dielectric matrix. These particles are loaded at a level and dispersed to an extent that they do not conduct on the plane of the adhesive tape.

Some of the Z-axis bonding agents use conductive particles in thermoplastic resin and are commonly known as hot melt adhesives. The particles size system is about 10–25 microns in diameter. Such adhesives require substantial pressure (3.5–20 psi) and temperature (130–150°C) to create a reliable contact. These Z-axis adhesives are commonly used as strips for bonding LCD's and flex circuits to PCB's. Many of the large LCD computer displays use Z-axis adhesive to mate flex driver circuits to the contacts on the edge of the glass panels.

Isotropic bonding agents can be made of both thermoplastic or thermosetting polymers. Conductive epoxies are the most common class of isotropic adhesives for component attachment. The isotropy comes from the fact that the conductive particles are highly packed to provide equal electrical conduction in every axis. Stencil printing is preferable to screen or pneumatic syringe dispense because better definition can be achieved. Resin bleed is also avoided. After application of adhesive, the components are placed onto the wet material followed by hardening or chemical curing. No flux is required for PTF assembly, making it an intrinsically clean technology.

10.4 ELECTRICAL PARAMETERS OF PTF CIRCUITS

10.4.1 DC Characteristics

The DC characteristics of PTF circuits are defined by their resistance and current capacities. The resistance measurement unit is in ohms, and current capacity is in amperes. The resistance of the circuit is measured at the circuits operating temperature (nominally 25°C). Whereas the current capacity of a conductor is defined as the maximum current that the conductor is capable of carrying without degrading of conductor, laminating adhesive, or dielectric. The glass transition temperature (T_g) is sometimes used as the maximum temperature for the dielectric, although many plastics have good characteristics above their T_g. Performance is also dependent on substrate thermal capacity. Because the traces are typically thin and wide, the heat dissipated is solely determined by the track surface area through natural convection and radiation. However, it was determined experimentally that convection is the dominant heat loss mechanism as compared to radiation in the PTF circuit.

Experimentally, it has been determined that the rate of heat loss from a thin wire in air due to natural convection is approximately given by Eq. 10.4.

$$\frac{1}{A}\frac{dQ}{dt} = 2.09 \times 10^{-4} \times \alpha \Delta T^{1.25} \text{Joules/cm}^2 \text{ sec} \tag{10-4}$$

where

dQ/dt = the rate of heat loss in Joules/sec
A = heat loss surface area in cm^2
ΔT = the temperature difference between the heated surface and the surrounding air
α = the empirical wire factor.

The radiation loss from a thin wire is given by the following Eq. 10.5.

$$\frac{1}{A}\frac{dQ}{dt} = 5.73 \times 10^{-12} \times \beta \, T_0^3 \, (T - T_0) \text{ Joules/cm}^2 \text{ sec} \tag{10-5}$$

where

T = the absolute temperature of the heat emitting body (in °K)
T_0 = the equivalent black body temperature of the surrounding environment in (°K)
β = the emissivity of the surface, which is the ratio of heat emitted by it to that emitted by a black body. This value has to be less than 1.0

the amount of heat loss. The heat generated in the PTF film is equal to the ohmic loss from the silver trace resistance. Because the conductivity of the silver track is given by Eq. 10.1, the corresponding resistance is given by Eq. 10.6

$$R = \frac{L}{\sigma_v wt} \Omega \tag{10-6}$$

The ohmic loss per unit surface area is given by Eq. 10.7:

$$\frac{1}{A}\frac{dQ}{dt} = \frac{I^2R}{A} = \frac{I^2 L}{\sigma_v tw\,(2Lw)} = \frac{R_\square}{2}\left(\frac{I}{w}\right)^2 \text{ Joules/cm}^2 \text{ sec} \tag{10-7}$$

Note that the effective surface area for heat loss is $2Lw$. Equations 10.4 and 10.7 show that a current versus temperature relationship is achieved. This is given in Eq. 10.8

$$I = \frac{1.430 \times 10^{-2}\ \alpha\ \Delta T^{0.625}\ w}{R_\square^{0.625}} \text{ Amps} \tag{10-8}$$

where $R_\square = 1/\sigma_v t$

Figure 10.6 shows the empirical current versus temperature on a PTF circuit

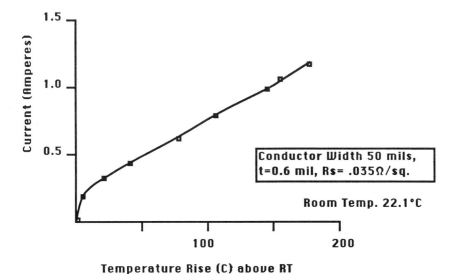

FIGURE 10.6   Current versus temperature

$w = 0.127$ cm (50 mils) width, $t = 0.00152$ cm(0.6 mils) thick with a $R_\square = 0.035$ W/s. An empirical average value of α can be obtained by matching Eq. 10.8 with the empirical data, and this value is found to be 3.1884 Amps. This is a reasonable value because a thin wire can be as high as 20 Amps.

If $\Delta T = T_g - T_{RT}$ where T_g is the glass transition temperature and T_{RT} is room temperature (22°C), the current at which the sample reaches T_g can be computed from Eq. 10.7. As an example, a T_g value of 70°C would require a current of about 0.53 amperes. Since local heating does occur, a maximum temperature of 60.4°C should be imposed. At this temperature, the current flow is about 0.80 Amp.

In order to understand that the radiation heat loss mechanism is insignificant as compared to the natural convection mechanism, a hypothetical calculation is given as follows. Assume the PTF circuit to rise from 22 to 100°C, it would be useful to determine the amount of heat loss in each case. Using Eq. 10.6, the amount of heat loss through radiation is 0.01147β joules/cm^2 sec. The equivalent current to maintain heat loss due to natural convection and maintain at 100°C based on Eq. 10.8 is 0.71652 amperes. The equivalent heat loss due to convection is 0.55704 joules/cm^2 sec. If the heat loss mechanism due to radiation is equal to the same amount as natural convection, it would demand the value of β to be about 48.56. As mentioned earlier, the maximum value for β, the emissivity of the surface can be 1.0. Even under such a condition, the heat loss due to radiation is less than 2%. In conclusion, radiation loss in PTF is negligible.

10.4.2 AC Characteristics

The AC parameters of PTF circuits are characterized by impedance, propagation delay, crosstalk, and signal attenuation. At high frequencies, these conductors start to exhibit wave behavior. At low frequencies, signal voltage relation is expressed using the simple Ohm's law, $V(x) = V(x = 0) - I(x = 0) R_{DC}$ (see Figure 10.7). This means that signal voltage at any point along the line is given by the signal voltage at $x = 0$, reduced by the voltage drop across the line, $I_s R_{DC}$, where $I(x = 0)$ is the signal source current, and R_{DC} is DC resistance of the line. This relation suggests that the signal voltage is attenuated by an amount equal to the DC voltage drop due to the current flow across the resistive line. The signal current is in phase with the source voltage over the line (assuming $I \ll \lambda/8$).

However, at high frequencies, where $I > \lambda/8$, the signal voltage $V(x)$ and current $I(x)$ variation over the line are significant and are given by:

$$V(x) = V(x = 0) \cosh P_x - I(x = 0) Z_0 \sinh P_x$$

$$I(x) = I(x = 0) \cosh P_x - (V(x = 0)/Z_0) \sinh P_x$$

where P_x is the propagation constant (complex number).

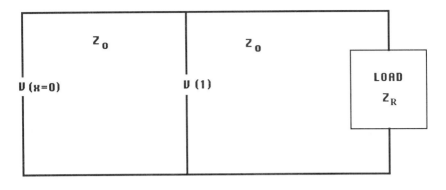

FIGURE 10.7   AC impedance

The real part of the propagation coefficient P_x determines the reduction in magnitude for the current or voltage along a line and is called the attenuation coefficient, α. The imaginary part of P_x is called the wavelength coefficient, β. The sending impedance of a line terminated with impedance, Z_R, is given by Eq. 10.9

$$Z(x = 0) = \frac{V(x = 0)}{I(x = 0)} = Z_0 \left(\frac{Z_R \cosh P_1 + Z_0 \sinh P_1}{Z_0 \cosh P_1 + Z_R \sinh P_1} \right) \qquad (10\text{-}9)$$

Equation 10.9 tells us that the impedance along the line seen by the source is no longer uniform and simple. If Z_R is equal to Z_0, then the impedance of the load is matched to the line, and the impedance is uniform throughout the line and does not have voltage or current reflections. Any voltage and current reflections can cause performance problems because a reflected pulse can cause a fast change in the logic state. It therefore limits the speed of the logic system.

Figure 10.8 illustrates the change of impedance seen by the signal over the frequency range of 100 kHz to 200 MHz for a PTF line having top track width, 100 mils and bottom track width, 200 mils. At 100 kHz, the impedance is about 60 Ω, which is equal to the 50 Ω load plus the DC resistance of the overall line. However, at frequencies above 40 MHz, it drops below 10 Ω. The phase shift changes from 0 to $-7°$ when signal frequency increased from 100 kHz to 40 MHz. Figure 10.8 illustrates the effect of a reflection in the frequency domain. A change in line width of the trace from 10 mils to 100 mils changes the line impedance. The dip in impedance to 50 Ω occurs at 132 MHz for a signal launched from narrow to wide line width and when the launching port is reversed, the frequency dip drifted to 139.5 MHz, and the impedance is reduced to 40.7 Ω. The distance between the line width change and the connector differs.

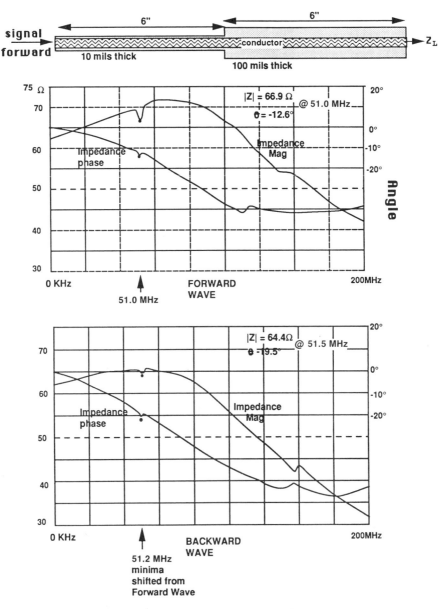

FIGURE 10.8   Impedance versus frequency

There is a change in the curve shape for the narrower conductor, seen as a rising hump, due to the higher impedance.

Besides the changes of impedance along the line, there are radiation and dielectric losses to encounter as frequency increases further. The conductive line can no longer confine the electromagnetic waves between the tracks and the dielectric. In addition, the dielectric no longer behaves as a perfect dielectric, and the insulating media actually permits leakage current to flow through. The track is no longer just resistive, it responds to the magnetic field setup, opposing the original alternating current. The track must be represented as an inductor in series with resistance. In the copper circuit, normally r is small compared to ωL, and G is small compared to ωC. The characteristic impedance of the line is the electrical wave impedance of the long conductive tracks. The characteristic impedance of a copper circuit can be given by Eq. 10.10.

$$Z_0 = \sqrt{\frac{r + j\omega L}{G + j\omega C}} \simeq \sqrt{\frac{L}{C}} \; \Omega \tag{10-10}$$

However, in a PTF circuit r is no longer insignificant when compared to ωL in a copper circuit. Equation 10.10 would have to account for r and would be frequency dependent. This is given by Eq. 10.11.

$$Z_{0PTF} \simeq \sqrt{\frac{r + j\omega L}{j\omega C}} = \sqrt{\frac{L}{C}} \sqrt{\left(1 - \frac{jr}{\omega L}\right)} \tag{10-11}$$

Equation 10.11 also indicates that the presence of r in the impedance imposes a complex quantity, thereby causing a phase shift between the sending voltage and sending current. This phase shift can prevent an oscillator from functioning in a circuit, such as the crystal oscillator circuit found in the microcontroller- or microprocessor-based system. The lossy property of the PTF circuits determines its dispersive nature.

Figure 10.9 shows an inductance coil made out of a PTF circuit, and it illustrates a relatively low parallel resonant Q value of 1.59 at $f = 6.04$ MHz and a wide bandwidth of $\Delta f = 6.2$ MHz. The corresponding phase shift is 58°C. The resistance value in Figure 10.8 is given by $r = 525.89$ Ω with an inductance of $L = 8.69$ mH. Using the approximate relation $\theta = \tan^{-1} r/2\pi f L = 58°$. The magnitude of the impedance, $|Z| = 279.2$ Ω.

Another parameter, propagation delay, is the time it takes a signal to propagate along the tracks. It is defined as the reciprocal of the signal velocity, i.e. $T_{pd} = 1/v$, and it has units of nanoseconds per inch. For example, a signal would take approximately 5.6 ns to travel a track length of 30 in. if the relative permittivity of the substrate material is 5. This propagation delay is found in all PCB.

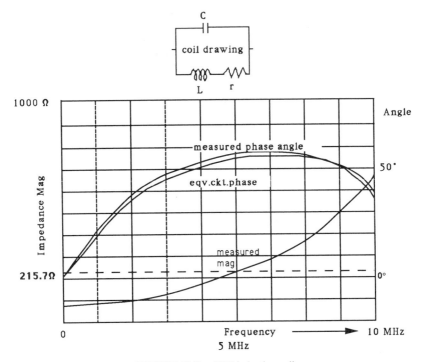

FIGURE 10.9   PTF induction coil

Crosstalk can occur both ways—signal lines crossing over one another and signal lines parallel to one another. Lines crossing over one another are usually treated as capacitive, but lines parallel to one another can have both capacitive and inductive coupling. In the parallel line cases, the crosstalk can be divided into forward and backward, depending on the direction of the wave travel with respect to the signal direction. The forward crosstalk coupling coefficient is a direct function of mutual capacitance, mutual inductance, and intrinsic line impedance. This constant K_{cf} is given by Eq. 10.12.

$$K_{cf} = \frac{1}{2}\left(C_m Z_0 - \frac{L_m}{Z_0}\right) \text{ nanoseconds/inch} \qquad (10\text{-}12)$$

where

C_m = the mutual capacitance
L_m = the mutual inductance
Z_0 = the line impedance

The amount of forward crosstalk voltage developed is equal to the sum of product of risetime of the signal and the coupling length of the line weighted by the crosstalk coupling coefficient. The backward crosstalk coupling coefficient is given by Eq. 10.13.

$$K_{cb} = \frac{1}{4T_u} \left(C_m Z_0 + \frac{L_m}{Z_0} \right) \qquad (10\text{-}13)$$

where T_u is the transit time of the back signal

It should be noted that the coupling coefficient, K_{cb}, is dimensionless. Therefore, in the AC analysis of the signal path of the PTF circuit, it is important to measure these parameters. The measurement for crosstalk in PTF circuits is made more difficult because of its high dispersive nature.

Because PTF circuits have high resistance as compared to copper circuits, they are highly dispersive. They require more accurate modeling. Accurate impedance measurements along each section of the PTF circuit can be achieved by using a Time Domain Reflectometry (TDR) instrument. Once a model of the PTF circuit is developed, a system performance analysis of the PTF circuit with the electronic circuit can be performed. A simulation of tolerance can be performed for the range in variation of component parameters. Such a simulation will indicate how tight a manufacture tolerance of the PTF circuit needs to be. Therefore, a highly dispersive PTF circuit requires more modeling than the regular copper circuit.

10.5 SURFACE MOUNT ASSEMBLY

In this section, the conventional solder Surface Mount Devices (SMD) process is first described and then a conductive adhesive attachment assembly is discussed. First, the PCB boards are inspected for shorts and opens, and the pick and place equipment is used to mount components. In the conventional Plated Through Hole circuit board, axial leaded components are used. These leads are placed through the holes and are commonly wave soldered from below the board. The advent of SMD allows components to be mounted on both sides of the circuit. These components have replaced their axial leads with contact pads.

The conventional assembly of the double sided surface mount assembly uses dot dispensed UV curable adhesive (i.e., pneumatic pressure to extrude adhesive from a syringe) to attach the components on the underside of the board before IR reflow or wave solder process. In a conventional wave soldering technique, through hole components assemblies pass over a wave (or two waves) of molten solder. Such a technique is particularly favored for mixed print assemblies where through hole components are on top, and SMD's are on the bottom. Wave soldering is undesirable for SMD assemblies because the presence of SMD's on the solder side of the board force the solder to go around and over the components instead of reaching the contact lead pads of components. This effect is generally known as shadowing, and it creates dry-joint. However, IR reflow is favored in an all SMD assembly. The tackiness of the nonconducting adhesive is of particular importance to keep the component in place before UV curing. Once cured the components are held in place for IR reflow. The solder paste or cream, which is composed of a fine suspension of solder particles in a sticky resin base flux, is first screen or stencil printed on the printed board. The board would then pass through an IR oven raised above 200°C to melt the solder paste which would coalesce together and wick up on the center of the component land pads. A cleaning process will then be initiated to remove harmful contaminants. Because SMD components are mounted much closer to the boards than axial leaded components, solvents have difficulty in flowing through the underside of the components. Cleaning techniques involve the use of CFC solvents, high pressure sprays, and water with saponifiers. The cleanliness is determined by measuring the conductance of a 75/25% by weight solution of isopropanol and water. These assembled boards will be inspected and tested for shorts, opens, and in-circuit board testing. Testing SMD assemblies differs from testing through hole assemblies because the device leads cannot be probed because probe pressure on device leads might press an open lead into contact and give a false pass. If there are assembly failures, rework to replace components must be done.

The SMD footprint design is important to include the bridging between solder pads and allow for the appropriate solder meniscus. Surface mount land patterns are published by the Institute for Interconnecting and Packaging Electronic Circuits (IPC), and it is found in the document ANSI/IPC-SM-782. These patterns help to optimize the strength of holding the components by its pads without solder bridges causing shorts.

In the polyester substrate copper flex circuits, the copper tracks are wide and numerous serving as a heat sink. Since the polyester substrate is sensitive to high temperature, the excess heat sinking property keeps the substrate protected. Both SMD and axial leaded components can be used in such an assembly because the flex circuit has through hole vias and also through hole component placement. These soldered assemblies also require cleaning.

In the conductive epoxy assembly process, the adhesive takes the place of solder. The adhesive is tacky enough to hold components in place after the pick and place operation. However, the IR reflow or the wave solder process is not required, and it is replaced by an IR curing process that is much lower in temperature than IR reflow or wave soldering process because the curing temperature is maintained below the substrate deformation temperature. It also takes a longer time, since curing of an epoxy requires a longer period than melting the solder. The one advantage of the epoxy attachment system is that a cleaning solvent is not required. This is favorable to the clean environmental issue because most cleaning solvents involve fluorocarbon compounds.

CONCLUSIONS

The PTF technology, used in flex circuits, is a direction towards low cost, lighter circuits. However, the material system used is vastly different from the conventional copper PCB. The choice of materials in PTF circuits, such as the conductive tracks, substrates, and dielectric insulator layer opens a much larger scope of materials compared to copper PCB and it also allows blending of one material system with another, such as carbon and silver particles. The lowering of a high resistivity parameter of the polymer conductor is still a challenging area for improvement. There is a future for use of different kinds of bonding agents for electronic components. The assembly of PTF circuits can lead the PCB industry into a simpler, more cost effective and environmentally cleaner circuits for the future. It also allows simple 3-D contouring of a planar circuit to fit most package housings.

References

1. Gilleo, K. et. al., "Screened Through Hole Interconnect Process With Plated PTF Ink," U.S. Patent 4,747,211, 1988.
2. Gilleo, K. Screened Through Hole Technology, *SITE*, pp.18–21, Feb. 1988.

11

Performance and Reliability

Joseph F. Fjelstad

J. F. Fjelstad Associates

11.1 INTRODUCTION

The performance and reliability of any product is often a significant measure of customer satisfaction. In the case of consumer products, a poor performing and unreliable product may merely be a nuisance, causing the purchasers to foreswear any future buying of shoddy products. In such a case, the nonperformance product is the executor of its own demise. However, in a great many cases, performance and reliability are paramount issues intrinsically linked to the success of the product. For example, we want the brakes on our automobiles to perform (stop the car), and we need for them to do so reliably (100% of the time). If the brakes do not conform to these requirements, an accident and possibly injury or even death can be expected to ensue.

These same concerns of performance and reliability are no less important in the manufacture of flexible circuits. Their importance cannot be trivialized, especially as flexible circuits become more and more pervasive in electronic packaging. Electronic packages that are used in every conceivable application from heart pacemakers, to automotive instrument clusters, to missile guidance systems.

If a flexible circuit does not perform in these applications, the ramifications are pretty grim.

Having established the importance of performance and reliability, how does one intelligently go about achieving it? One way would be to build the product out of all the different available materials and in all conceivable ways and then devise appropriate tests, and test them all to failure to determine the best total system. The number of perturbations would be unwieldy in many cases. Another way to create a reliable product is to draw from the experiences of others and fabricate that product using the best available standards and specifications of design and performance. A reasonable person would probably choose the latter approach over the former.

The purpose of this chapter is to introduce the reader to the fundamentals of performance and reliability of flex circuits by providing them with background on the various standards and specifications that are employed by industry and further to discuss the testing and test methods that are used to ensure that the specifications are being met.

11.2 STANDARDS AND SPECIFICATIONS

Standards and specifications are basically performance criteria documents that can be generated in two ways. First they can be generated by product end users who would use the specifications to describe, in very specific terms, how they want their products to perform when subjected to testing. The second way is by the product vendors, who use product specifications to describe what their product can do when subjected to appropriate tests.

Obvious problems arise when these two methods produce conflicting results. It is these conflicts that were responsible, at least in part, for the development of all industrial and military standards to specifications and the agencies that act as curators for these documents. Industry documents are consensus statements reflecting, in the most realistic fashion possible, the levels of performance that can be expected from a product as agreed to by both manufacturer and user. As such, industry specifications can save product users and vendors much time and expense. They are readily adaptable to serve immediate mutual needs for virtually all circuitry products. It is important to remember that they do not pre-empt an individual user from demanding greater levels of performance in any or all areas of the specification, but at least they can start from common ground and from a known experience base. Remember, also, that the extra demands will normally be met with demands of a manufacturing premium, especially if the new requirements are beyond the range of present experience.

The difference between standards and specifications is subject to occasional confusion. In simplest terms a *standard helps to define how something is to be done*, and a *specification defines how it is to perform*. Stated another way, design standards are created to guide the design of a product so that it can meet the appropriate specification.

In summary, standards and specifications are vehicles for communication between user and vendor through which an agreement on the performance requirements of the end product can be sought.

11.2.1 Standard and Specification Generation Agencies

There are numerous specification generating agencies around the world. Some are government sponsored, and many others are industry sponsored. Most often, the industry that creates a certain type of product also has a trade association that helps the industry generate its standards and specification and acts as curator for them. For flexible circuitry, there are several groups around the world who have generated specifications covering their performance. For purposes of classification, the author has divided them into three very broad and often overlapping groups: (1) Industrial, (2) Europe and Asia, and (3) U.S. Government, Military and Aerospace.

Industrial Agencies.   *United States.* Industrial agencies are normally trade associations, confederations of manufacturers and often the manufacturer's customers and vendors as well, who jointly develop specifications for their product(s). Many industrial agencies participate in the development of flex circuit standards. A few of the more prominent ones are the following:

1. IPC—Institute for Interconnecting and Packaging Electronic Circuits. Located in Lincolnwood, Illinois, the IPC (originally Institute of Printed Circuits) was founded in 1957 to create a forum for standards development of the fledgling printed circuit industry. It has since that time expanded in scope to embrace the entire body of interconnections between the silicon chip and the electric wall socket.

The IPC has been responsible for generating the most prominent specifications for flexible circuits and is international in influence.

2. ANSI—American National Standards Institute.   Founded in 1918 and headquartered in New York, N.Y., ANSI is a national coordinating body that is a clearing house for standards and specifications for all types. ANSI serves to help prevent redundancy and conflict in specifications and provides American National Standards status to standards and specifications developed by agreement of all concerned groups.

3. NEMA—National Electrical Manufacturers Association.   Founded in 1926 with headquarters in Washington, D.C., NEMA serves, in part, to develop standards and specifications for nomenclature, performance, and testing on a broad range of products, all of which are concerned with either the generation, transmission, distribution, and control or use of electrical power. NEMA is quite familiar to most PWB manufacturers because of ratings developed for circuit board laminates by its insulating materials division.

4. EIA—Electronic Industries Association.   Founded in 1924 with its main office in Washington, D.C., the EIA is a large and multifaceted association that

covers a broad range of electronic related disciplines. The EIA is also cosponsor to many developed standards and specifications.

Europe And Asia.   1. EIPC—European Institute of Printed Circuits. Founded in 1968 and based in Lugano, Switzerland, the EIPC is more a forum for the exchange of information on circuit technology than a standards agency. Its members, however, do participate in the review of other agency's standards and specifications.

2. ICT—Institute of Circuit Technology.   Founded in 1973 with offices in Oxford, England, ICT also promotes information exchange and represents members' interest before international standards and specification committees.

3. IEC—International Electrotechnical Commission.   In operation since 1906 and based in Geneva, Switzerland, the IEC is the grandfather of standards organizations. The IEC, through its many committees, seeks to establish international standards for electrical and electronic products among its 42 member nations.

4. AEU—Asia Electronics Union.   Established in 1968 in Tokyo, Japan, the AEU was formed to exchange information and promote cooperation among its 15 member nations of Asia and Oceania.

U.S. Government, Military and Aerospace.   There are several government and military agencies that generate specifications. Within the military there are also different groups. Conflict between general specification bodies, such as DESC (Defense Electronics Supply Command), the DOD (Department of Defense), and the tri-services of the Army, Navy, and Air Force have frequently occurred. In addition, NASA (National Aeronautics and Space Administration) and NSA (National Security Agency) have also developed standards and specifications. However, for purposes of discussions on flexible circuits, there is one main impacting agency and that is DESC, located in Dayton, Ohio. DESC is the curator of the military specifications on electronics and electronic assemblies. Under recent directives, DESC is also seeking to remove itself whenever possible from its specification generating role. Working closely with industry agencies, such as the IPC, an increasing number of industry specifications are being adapted and adopted by DESC, thus eliminating needless duplication of efforts.

This has been accomplished largely through the establishment of a class system within industry documents that recognizes different levels of performance are required for different products (i.e., the performance of a toy is of lesser importance than the performance of a life support system).

This cooperation between industry and the military is a welcome occurrence and should help to greatly clarify the matters of performance requirements now and in the future.

11.2.2 Specifications and Standards for Flex Circuits

Following are descriptions of some of the most commonly cited specifications and standards for flexible circuits. For purposes of brevity, it is here stated that virtually all IPC standards and specifications contain the following common elements:

1. They provide a measure of scope giving definition to the limits of the document's subject of interest.
2. They furnish a method for designating the various materials or products which are represented under the general title of the document.
3. They provide a method for distinguishing between the different classes of end use products in which the material or product being specified may ultimately be used (this concept was adopted to support the fact that the requirements for a life support system are vastly more critical than those for a child's toy).
4. They provide references to other applicable documents.
5. They give definition to the requirements of the material or product, including visual, dimensional, mechanical, chemical, electrical, and environmental.
6. They set forth quality assurance provisions, including information on sampling plans and inspection.
7. They provide test methods for verifying product quality conformance.
8. Finally, when appropriate, they furnish slash sheets, which give detailed information on the performance limits of individual materials and products.

IPC-A-600 Acceptance of Printed Wiring Boards.   One of the oldest of IPC documents, the IPC-A-600 provides visual standards for acceptance of printed wiring boards and includes a special section highlighting flexible circuits. Because many of the criteria for rigid boards and flex circuits are the same, the rigid board information can also prove of value.

IPC-CF-150 Copper Foil Specification for Printed Wiring Application.   This document specifies the eight classes of copper foil that are presently available for printed wiring applications. The specification also describes, in detail, acceptance criteria and inspection and testing requirements for the copper foil.

IPC-FC-231 Flexible Bare Dielectric for Use in Flexible Printed Wiring.   This IPC document defines classes and specifies the requirements for flexible base materials that are used in the manufacture of flex circuits.

IPC-FC-232 Adhesive Coated Films For Use as Cover Sheets for Flexible Printed Wiring.   This specification details the requirements for those adhesive coated dielectric films that are coated on one side only and destined for use as coverlayers for flexible circuits.

IPC-FC-233 Flexible Adhesive Bonding Films.   This specification provides the requirements for flexible dielectric films that are coated on one or both sides with an adhesive and also for supported and unsupported adhesive films to be used in the fabrication of flexible and rigid flexible printed wiring.

IPC-FC-241 Flexible Metal Clad Dielectrics For Use in Fabrication Of Flexible Printed Wiring.   This specification establishes the requirements for flexible metal clad dielectrics that are used in the fabrication of flexible and rigid flex printed wiring.

IPC-RF-245 Performance Specification For Rigid Flex Printed Boards.   This document provides performance specifications and qualification tests for rigid flex printed wiring boards.

IPC-D-249 Design Standard For Flexible Single- and Double-Sided Printed Boards.   This document covers the various requirements and considerations that need to be addressed when designing single- and double-sided flexible circuits. The document is particularly valuable in that it is based on the experience of industrial manufacturers and their knowledge of common manufacturing capabilities.

IPC-FC-250 Performance Specifications For Single- and Double-Sided Flexible Printed Wiring.   This specification, a companion of sorts to IPC-D-249 provides performance specifications and qualification requirements for flexible one- and two-sided printed wiring boards.

MIL-STD-2118 Flexible and Rigid Flex Printed Wiring For Electronic Equipment Design Requirements.   This document provides pertinent design information required to facilitate successful manufacture of flex and rigid flex boards destined for military consumption.

MIL-P-50884 Printed Wiring, Flexible and Rigid Flex.   This specification establishes the performance criteria for flexible and rigid flex printed wiring boards that are to be used in military electronic equipment.

11.3 TESTING AND TEST METHODS

11.3.1 Purpose and Philosophy of Testing

The purpose of testing is obvious. Tests are performed to make certain that a product is capable of meeting certain criteria. It makes no difference if the product is a computer chip, an automobile, or a high school student. Without some sort of testing, we have nothing to measure the product with and thus have no certainty about how it will perform now or in the future. The product determines how to go about testing and the level of testing required.

One could not very well test a single student in a class room and say that they represented the whole class, even though they all received instruction from the same teacher. The individual students are influenced and affected by too many other factors during the other 23 hours of the day and have too vast an array of aptitudes and backgrounds to consider them a homogeneous population, thus they must be tested individually.

Fortunately, for most manufacturers, the complexity of their products does not remotely approach the complexity of a human being (though arguments could certainly be advanced that some of NASA's space programs exceed human complexity), and testing is a simpler task. In fact, a somewhat reduced level of testing is not only possible but imperative if economically viable products are a goal.

Testing, performed economically, is normally done on a sample basis. The sample product submitted to a testing program assumes the role of a typical product. Manufactured by the same process line, all product off the line should be, for all intents and purposes identical, subject to the condition that the process is in a state of control. Variance in the process is monitored, and testing is performed on a sample basis to verify that the statistically controlled process, still, in fact, yields an acceptable product.

Testing, by the way, should not be confused with inspection. Inspection is too often a sorting process for separating the good from the bad and the possibly bad. With respect to flex circuits, there are many areas where testing is performed to assure end product reliability. The following general areas are all in need of discussion.

Raw Materials. Raw materials are tested to assure that the end product will not be limited in performance by nonconforming basic construction materials. To enhance the users sense of reliability, raw materials (base materials, adhesives, and laminates) are tested in the following categories: Physical Performance, Chemical Performance, Electrical Performance, and Environmental Performance.

Physical Performance: To determine physical performance, the raw material is checked to ascertain that certain minimum physical properties can be achieved including:

- Tensile strength and elongation—This testing is performed to verify the material's physical toughness.
- Initiation tear strength—This test provides a determination of how much force is required to initiate tearing in a material.
- Propagation tear strength—This check provides a measure of force required to propagate a tear in a material.
- Low temperature flexibility—This test is performed to check material resistance to fracture at low temperatures.
- Dimensional stability—Dimensional stability is checked to verify that the material does not shrink or stretch excessively as a result of thermal exposure (or removal of metal in the case of clad laminates).
- Peel strength—Peel strength is measured under three separate sets of conditions. 1) As received to check for minimum bond strength of copper to the flexible dielectric base. 2) After 10 seconds 550°F solder float to check for degradation after simulated assembly and rework thermal stresses. 3) After aging or exposure to five cycles of $-50°C$ $+150°C$ with one-half hour dwells at each extreme with a 15-minute intermediate dwell at room temperature between exposures at extremes. Simulated accelerated aging effects bond strength.
- Volatile content—Volatile content is examined to determine the percent of volatility in flexible laminates.

Chemical Performance

- Chemical resistance—Chemical resistance is checked to verify that no deleterious effects occur to the laminate during exposure to commonly used processing chemicals.
- Flammability—This test is performed to determine what minimum levels of oxygen are required to sustain combustion of the laminate.

Electrical Performance: Key electrical performance criteria are verified by submitting samples to the following tests:

- Dielectric constant—This testing is performed to ascertain the value for the dielectric constant of the material. This is an important factor in many electronic calculations. It is simply a ratio of the capacitance of the material as opposed to the capacitance of air.
- Dissipation factor—This value is also determined to facilitate calculations and predictions of the performance of electronic circuits fabricated from the chosen material. The dissipation factor is the relationship between the materials permittivity (capacitance) and its conductivity, measured at a given frequency.

- Dielectric strength—This value is obtained in order to know the minimum breakdown voltage of the material. It is of particular importance in the fabrication of high voltage circuits and for circuits that are to operate at high altitudes.
- Insulation resistance—The insulation resistance of a raw material is checked to provide base line data as to what maximum resistance values can be expected from a flex laminate.
- Surface and volume resistivity—These values are obtained to measure the volume and surface electrical resistance of the base material under moist conditions, such as might be encountered during product life.

Environmental Performance: Following are tests performed to evaluate certain circuit attributes that may change as a result of exposure to the environment.

- Moisture absorption—Measured to determine how much moisture the base material can absorb. This is also an important factor for electrical considerations, as moisture absorption can affect the dielectric constant of the material.
- Moisture and insulation resistance—This value is measured in opposition to insulation resistance to determine the effects of moisture exposure on the insulation resistance of the raw material.
- Fungus resistance—This test is performed to determine if the material is adversely affected by the presence of fungi under conditions favorable to their development.

Flexible Circuit Testing.   Beyond testing of flexible circuit raw materials, there is also the need to test the circuits themselves. Many of the tests are duplicated to assure that the manufacturing process has not degraded on the material beyond acceptable limits. Requirements are established in several key areas to which the product must comply in order to be acceptable for its class.

Major test areas include: visual, solderability, dimensional, physical, construction integrity, plated through hole after stress, electrical, environmental, and cleanliness.

Visual Requirements: Visual inspections are performed to assure conformance to minimum requirements in the following areas:

- Delamination—An indicator of improper processing or thermal stress excesses. In nearly all cases, delamination is viewed as a rejectable condition.
- Edge condition—Flex circuits are inspected to check for nicks or tears that could be sites for tear propagation and failure of the circuit.

- Solder wicking—Evidence of solder wicking along the surfaces of traces away from openings in the cover layer may be an indication of inadequate cover layer lamination or excessive exposure to molten solder. The condition is not preferred but is acceptable, if within specified limits.
- Stiffener attachment—Because stiffeners perform vital mechanical support functions, the attachment is checked to assure integrity of the attachment method.
- Plated through hole voids—Inspected for the preempt soldering or reliability problems due to voids in the copper of the plated through hole.
- Marking—Marking is checked to verify compliance to the master drawing and prevent mislocating or misorienting of components in assembly or repair.
- Workmanship—A general category inspection that covers miscellaneous items, such as presence or absence of finger prints, dirt, and other non-specific indicators of the workmanship used in the manufacture of the circuit. While somewhat subjective, it often provides an indication of the overall quality of the circuit.

Solderability Requirements: Solderability is tested to ascertain that the flexible circuit will not experience or create problems in assembly operations. This is normally accomplished by checking to make certain that proper wetting of solderable surfaces is present.

Dimensional Requirements: Dimensional inspections are performed to assure the fit and form of the circuit. They are also checked because dimensional accuracy is often important to the function of the circuit. Following are some key checks.

- Hole pattern accuracy—Checked to assure that holes are in their proper locations. This is an assembly related issue and is of most importance when automatic assembly procedures are employed.
- Annular ring—Insufficient annular ring can effect both the solderability and the reliability of an interconnect. Requirements differ significantly for plated through and nonplated through holes and must be checked to assure minimum requirements are met.
- Cover layer registration—Similar to annular ring in net effect, the cover layer must register to requirements. Adhesive squeeze-out onto lands from under cover layer openings and screen printed cover coat misregistration are included in this requirement.
- Hole sizes—If too small, circuit hole sizes may make component insertion difficult. Excessively large holes may make it difficult to form preferred solder joints in the assembly process. Thus, hole sizes are checked preempting potential problems.

- Conductor pattern—The conductor pattern should accurately represent the master drawing. While allowances exist for localized reductions of up to 20%, excessive faults may be indicative of problems in manufacture.
- Conductor width and spacing—Conductor width and spacing must meet minimum requirements to assure proper electrical and/or electronic performance. This is normally controlled by the master drawing.

Physical Requirements: This area of testing is where the greatest departure from standard rigid board testing is found. It is also an area of testing that is of great importance in assuring product quality and reliability of flexible circuits.

- Plating adhesion—Plating adhesion tests are performed to assure a good metallurgical bond of plated metals to the laminate foil. Low bond strength could result in latent failure of the circuit.
- Unsupported hole bond strength—Unsupported hole bond strength is examined to assure that the circuit can endure assembly and repair without excessive damage.
- Conductor pattern bond strength—This test, in essence, retests the peel strength of the raw material, assuring that the circuit manufacturing process has not reduced foil bond strength to unacceptable levels.
- Folding Flexibility—As most flex circuits are mechanically formed or shaped at some time in their life, this test is performed to assure that they can be successfully formed without delaminating or breaking conductors.
- Flexibility endurance—This test is most important for dynamic flex applications as opposed to static applications. Standard test methods may require only minutes or hours to perform. Whereas, certain other test methods, such as those used in testing circuits for disk drive applications, may take several months.

Construction Integrity Requirements: The construction integrity of a flex circuit can best be evaluated only by cross-sectioning the product and evaluating it under a microscope. This type of inspection is used to reveal the presence of microscopic defects, which may impact the reliability of the finished flex circuit.

- Lifted land (before stress)—Check to assure no excessive land life is present before stress testing. Land lifting is normally the result of thermal stress, where the expansion of the heated laminate causes lands to lift from the surface of the circuit.
- Plating integrity—This check is performed to assure that plating quality, uniformity, and thickness meet requirements and that no cracks are present in the through hole. These anomalies, if unchecked, could impact the reliability of the circuit.

Plated Through Hole After Stress Requirements: The following checks are performed after different forms of thermal stress have been applied to the flex boards. Microsection evaluation of plated through hole cross sections is the approved method for determining hole quality.

- Thermal stress—This test is performed to determine if the plated through holes can successfully survive assembly. The test is actually much more punishing than normal assembly, but is used to cull out marginal product.
- Rework simulation—Soldering and desoldering of a lead in the plated through hole is performed to simulate field repair effects on the plated through hole's integrity.
- Lifted lands (after stress testing)—Though lifted lands are not uncommon after testing, limits are established to assure that lands do not come off in the field.

Electrical Requirements: Electrical requirements are specified to assure proper electrical performance of the flexible circuit. While continuity and shorting an insulation resistance are the tests most often called out for circuit acceptance, other tests may be required as necessary. Time domain reflectometry (TDR) testing of controlled impedance circuits is an example of such a test, if required.

- Circuit continuity and shorting—Tested to make certain that no opens or shorts are present in the board. This testing will assure conformance to design and, presumably proper circuit performance.
- Insulation resistance—This testing is performed to assure processing has not degraded the insulation resistance of the material beyond requirements.

Environmental Requirements: Environmental tests are performed to simulate, in a practical fashion, the effects of various environmental conditions on flex circuit quality and performance.

- Moisture—In this test, also sometimes referred to as moisture and insulation resistance testing, circuits are cyclically exposed to warm moist air to ascertain that no deleterious effects occur in the laminate and to assure that undesirable degradation of the insulation resistance of the flex circuit has not occurred.
- Thermal shock—This test is performed to determine the effect, if any, on the flex circuit as it is cycled from -65 to $+125°C$. The object of testing is to simulate, in accelerated fashion, the effect of thermal cycling on the quality and performance of the flex circuit.

Cleaning Requirements: These checks are performed to assure that the circuits do not carry an excessive amount of ionic or organic contaminates with them, thus having a potentially negative effect on assembly, long term reliability, or both.

- Ionic (solvent extract resistivity)—This test is performed to determine the level of ionic contaminates left on the board by washing them from the board, collecting the washing solvent, and measuring its change in resistance.

As you can see, there are many tests and checks required to assure the reliability of a flexible circuit. While arguments are often made that such tests do not add value but merely add cost, those same arguments can be muted by pointing out that a failure (and its attending ramifications) that occurs because testing was omitted could, under appropriate circumstances, prove extremely damaging. It is also important to note that not all tests are required for every lot as testing is normally split into two levels; those required for qualification and those required for acceptance. Acceptance testing is more limited in scope, thus reducing testing requirements. Finally, the reader is reminded that an effectively run manufacturing operation—one that rigorously employs statistical process control methods—may be able to significantly reduce the formal inspection operations by keeping the processes under constant control, thus achieving desired levels of performance and reliability at no extra cost. This methodology is, in fact, quickly becoming a customer requirement from an increasing number of flex circuit users.

Glossary

Accordion an electrical connector contact on which a flat spring is given a "Z" shape to permit high deflection without overstress.

Active Component any electronic component or circuit which adds amplification or has a directional function, such as transistors and integrated circuits (IC).

Active Device see Active Component.

Additive Plating deposition of copper or other metal, using electroless or electrolytic plating processes. This process is sometimes used to manufacture printed circuits.

Additive Process in circuit manufacturing, a method of producing conductor patterns where material is added to the bare substrate by printing or some other selective deposition process. See Subtractive Process.

Adhesion property of two or more surfaces bonding to each other.

Adhesive(s) a wide range of materials, including animal and vegetable type glues, rubbers, elastomers, thermosetting and thermoplastic resins, ceramics, and hot melts. Adhesives are used extensively for bonding, sealing and joining laminates, films foils, coils, conductors, etc.

Adjacent Conductor an electrical conductor next to another conductor, either in the same multiconductor layer or in adjacent layers. The term applies to printed circuits and flexible cables.

Alloy a combination of two or more elements, of which at least one is a metal. Alloys generally have different properties from those exhibited by their constituent elements. The term is also applied to plastics. Solder is an alloy of tin and lead.

Alternating Current AC; electrical current which periodically swings positive and negative. House current is AC.

American National Standards Institute ANSI; the U.S. government organization (made up of over 1000 trade organizations, societies, etc.) responsible for the development and promulgation of (among others) data processing standards.

Amperage the amount of electrical current, measured in amperes (A) or milliamps (mA).

Ampere A; a unit of electric current; current flowing through 1 ohm at 1 volt.

Anchor Spurs extension tabs on the conductors on printed circuits, especially flexible, to improve adhesion of the conductor as coverlay is applied over the tabs.

Angle Connector an electrical connector that joins two conductors end to end at a specified angle.

Anisotropic Conductive Adhesive a material that conducts electrical current in a single axis. Materials that conduct in the Z or vertical direction, can be used to interconnect circuits to flexible circuits or to components.

Annealing a process of holding a material at a temperature near, but below, its melting point to relax or remove stress but without distorting the original shape. The term is used extensively in the metal industries but may be applied to plastics as well.

Annular Conductor a number of wires stranded in three reversed concentric layers around a core. A high durability electrical conductor cable.

Annular Ring that portion of a conductor in an electronic circuit completely surrounding a hole in the circuit board.

ANSI American National Standards Institute; see full term.

Antistat antistatic agent, additive, coating or treatment applied to nonconductive materials that will conduct away static electricity to a ground. Antistats are used where static electricity can cause discomfort or harm to persons or where static discharge will destroy equipment.

Antistatic Electrostatic Discharge (ESD) protective material which resists electric charging. The material conducts or ''bleeds'' off the electrical charge preventing spark discharge.

AOI Automatic Optical Inspection. see full term.

APT Automatically Programmable Tools; see full term.

Area Array Tab ATAB; a circuit-like device for connecting individual silicon electronic chips to circuit boards. A form of Tape Automated Bonding (TAB) where edge locating pads and additional pads on the inner surface of an electronic chip are bonded together.

Artwork an accurately scaled configuration or pattern which is used to produce the artwork master or production master. Artwork is used in the imaging process, such as printing or photo reproduction.

Aspect Ratio in printed resistors, the ratio between the length of a film resistor and its width. Electrical resistance increases with length and decreases with width, thus the aspect ratio determines its value.

Assembly a group of subassemblies and/or parts that are put together; the total unit constitutes a major subdivision of the final product. When two or more components or subassemblies are put together by the application of labor or machine hours, it is called an assembly. An assembly may be an end item or a component of a higher level assembly.

ASTM American Society for Testing and Materials.

ATAB Area Array Tape Automated Bonding; see full term.

ATE Automated Test Equipment; see full term.

Attack Angle in screen printing, the angle between the squeegee face and the plane of the screen. Screen printing is used extensively in graphic arts but also in the printed circuit industry.

Automated Test Equipment ATE; computer controlled equipment able to test printed circuit boards or full electronic assemblies and systems.

Automatic Optical Inspection AOI; an automatic system using a scanning inspection video input device. Information is digitized, compared in a computer, and defective parts are marked or sorted. A pattern recognition or a Design Checking Rule (DCR) approach may be used. The method can be used to inspect printed circuits or other complex fabrications.

Automatic Test Generation ATG; computer generation of a test program based solely on circuit topology. This concept is used with Automated Test Equipment (ATE). See also Automated Test Equipment.

Automatically Programmed Tools APT; a program language for numerical control machines which is used primarily to generate N/C tapes used to program these machines.

AWG American Wire Gauge; a wire thickness standard.

Axial Lead leads coming from the ends of an electronic device; along the central axis. This type of electronic packaging is typically used for resistors and diodes.

Backbonding attaching an electronic chip to substrate by its back, leaving the face up.

Back Mounted when a connector is mounted from the inside of a panel or box with its mounting flanges inside the equipment.

Backplane Panel an interconnection panel into which PC cards or other panels can be plugged, ranging from a PC motherboard to individual connectors mounted in a metal frame. Panels lend themselves to automated wiring.

Ball Bond thermocompression bond formed when a ball shaped end interconnecting wire is deformed against a metal pad. Used to connect electronic ''chips.''

Barrel the portion of a terminal or contact that is crimped. There are two basic types: wire barrel which receives the conductor and the insulation barrel which grips an insulated post.

Barrier Strip a continuous section of dielectric material which insulates electrical circuits from each other or from ground.

Base Metal metal from which the connector, contact, or other metal accessory is made and on which one or more metals or coatings may be deposited.

Base Ply the outermost layer of a membrane switch.

Bayonet Coupling a quick-coupling device for plug and receptacle connectors, used by rotating a cam operating device designed to bring the connector halves together.

Beam Lead a long structural member unsupported along some of its length and subject to forces of flexure, one end of which is permanently attached to a chip device, and the other end intended to be bonded to another material for electrical interconnection and/or mechanical support.

Beam-Lead Tape in tape automated bonding, the kind of tape where the inner leads are formed as beam leads into an etched conductor pattern to contact the die pads. A method of connecting electronic ''chips.''

Beam-Lead Tape Bonding a cost-efficient, mass gang-bonding interconnection technique involving use of a continuous, sprocketed beam-lead tape for sequential automated bonding of its precision pre-positioned lead patterns.

Bed of Nails test fixture made up of base frame holding an array of spring loaded pins which make electrical contact with specific points on a printed circuit.

Bellows Contact a connector contact which is a flat spring folded to provide a uniform spring rate over the full tolerance range of the mating unit.

Bifurcated Contact a connector contact (usually a flat spring) slotted lengthwise to provide additional, independently operating points of contact.

Binding Post a fixed support, generally screw-type, to which conductors are connected.

Birdseye a fastener used to attach a flexible circuit to a hardboard. See Eyelet.

BIST Built In Self Test; see full term.

Blade Contact used in multiple contact connectors, a flat male contact which mates with a tuning fork or flat formed female contact.

Blank to cut to shape from a sheet or roll film with a punch or die.

Bleed (1) (printing) a condition where screen printing ink flows out beyond the original print pattern due to low viscosity or poor ink rheology. Also, migration of one color of ink into another. (2) (circuit) the condition in which a plated hole discharges process material or solution from voids and crevices.

Bleeding see Bleed.

Blister a dome shaped defect caused by a loss of adhesion between the deposit and the substrate or base metal and foil. Often due to outgassing.

Block (1) unwanted adhesion between stacked sheets often due to undercuring of ink or coating. (2) connector housing, usually made of molded plastic.

Bloom (1) (plastics) a surface exudate appearing as an oily or greasy film on a plastic surface. (2) (metallics) a haziness caused by color moving to the surface of a film. (3) (vacuum metallizing) a haziness or translucency often due to substrate plasticizer migrating through the thin metal film or by mold release or other contamination.

Bolted-Type Connector a connector in which contact between the conductor and the connector is made by pressure from clamping bolts.

Bond an interconnection which performs an electrical and/or a mechanical function in an electrical/electronic circuit.

Bond Deformation the physical change in a bond lead produced by the bonding tool as in wire bonding.

Bond Shear Strength it is to be calculated as the limiting stress, measured as a true shear force (in grams or Newtons) applied uniformly along the complete length of an item's side or contour, moving parallel to the bond interface and divided by the sheared-off bond surface area. Applicable to shearing a substrate from a package bottom, or die or device from a substrate surface.

Bond Strength the force per unit area or unit width required to separate two bonded layers by a force perpendicular to the surface.

Bonded Assembly a connector assembly with components bonded together with adhesive in a sandwich-like structure to provide sealing against moisture and other environments which weaken electrical insulating properties.

Bonding Island see Bonding Pad.

Bonding Pad a metallized area at the end of a thin metallic strip to which a connection is made. Also called Bonding Island.

Boot a form placed around wire termination of a multiple-contact connector to contain the liquid potting compound before it hardens. Also, a protective connector housing made from a resilient material to prevent entry of moisture.

Box Style Wire Contact a terminal strip design feature in which wire is completely enclosed in a contact and cannot be pushed through the connector.

Braid woven bare metallic or tinned copper wire used as shielding for wires and cables. Also a woven fibrous protective outer covering over a conductor or cable.

Braze to join metals using a non-ferrous filler metal at temperatures above 800°F. Sometimes called hard soldering.

Brazed Terminal solderless terminal with a barrel seam brazed to form one piece.

Breadboard a circuit simulation using actual components on a plug-in board.

Breakaway the off contact distance between the surface of a substrate and the bottom of a screen when screen printing.

Breakdown Voltage (1) the voltage threshold beyond which there is a distinct increase in electrical conductivity. Applies to dielectrics and semiconductors. (2) the voltage at which an insulator or dielectric ruptures or ionization and conduction take place in a gas or vapor.

Breakout the point at which conductors separate from a multi-conductor cable to complete circuits at other points.

Bridging (1) a defect condition, where the localized separation between any two conductor lines (paths, traces) is reduced to less than the minimum allowable. In the extreme, it becomes a short. Bridging may be caused by misalignment, screening, solder splash, smears, or attached foreign material. (2) a soldering defect wherein solder forms a shorting bridge between adjacent leads or conductors. (3) electrical— the formation of a conductive path between conductors causing a short circuit defect, often due to metal migration.

BTAB Bumped Tape Automated Bonding; see full term.

Buckle a crumpling occurring while a web is winding up sometimes seen in roll-to-roll circuit processing and in roll laminates.

Built In Self Test BIST; electronic circuitry with diagnostics.

Bulkhead connector designed for insertion into a panel cutout from the component side.

Bumped Tab a form of Tape Automated Bonding (TAB) substrate with small bumps of metal formed on its inner leads to ensure good bonding to electronic chips. This form of TAB is required unless the electronic chips have bumps.

Bumped Tape a tape for the TAB process where the inner-lead bond sites have been formed into raised metal bumps to ensure mechanical and electrical separation between ILB's and the non-pad areas of the chip (die) being bonded.

Bumped Tape Automated Bonding BTAB; bonding bumps to facilitate connection to the IC chip are formed on the TAB.

Buried Via an interstitial via hole not extending to the surface. See Via.

Bus (1) one or more electrical conductors that are used to transmit power or data in electronic circuits. (2) a circuit over which data or power is transmitted. (3) in computers, a common set of electrical pathways for signal transmission.

Butt Connector a connector in which two conductors come together end-to-end, but do not overlap with their axes in line.

Butt Contact a mating contact configuration in which the mating surfaces engage end-to-end without overlap and with their axes in line. This engagement is usually under spring pressure.

Butt Splice device for joining conductors by butting them end-to-end.

Button-Hook Contact a contact with a curved, hooklike termination often located at the rear of hermetic headers to facilitate lead soldering or desoldering.

C and W Chip-and-Wire. see full term and also Wire Bonding.

Cable Assembly a cable with plugs or connectors on each end.

Cable Clamp a device used to give mechanical support to the wire bundle or cable at the rear of a plug or receptacle.

Cable Terminal a device which seals the end of a cable and provides insulated egress for the conductors.

CAD Computer Aided Design; see full term.

CAD/CAM the integration of computer aided design and computer aided manufacturing to achieve automation from design through manufacturing.

CAE Computer Aided Engineering; see full term.

CAE-to-ATE Computer Aided Engineering-to-Automatic Test Equipment; an advanced computer system where electronic test equipment can be directly programmed from engineering work stations.

CAM (1) Computer Aided Manufacturing; see full term. (2) Content Addressable Memory; see full term.

Camber the overall warpage present in a substrate.

Capacitance the property of an electronic circuit or device to store electrical energy by means of an electrostatic field (stored electrons).

Capacitive Coupling the electrical interaction between two conductors caused by the capacitance between them.

Card Edge Connector a connector that mates with printed wiring leads running to the edge of a printed circuit board. Also called Edgeboard Connector.

Carlson Register Pin CRP; a low profile, circular registration pin on a thin, flat metal base that can be taped or bonded to machinery.

Carrier Web a web or film underneath the printed circuit web used for additional support or to transport sheet circuits through a roll-to-roll process.

Catalyst a substance, usually a chemical, which promotes or accelerates a process or reaction.

Caul a sheet of material, often with release properties, used in cold or hot, pressing of assemblies being bonded.

Cermet a solid homogeneous material consisting of powdered ceramic and metal. Cermet is used for conductors in ceramic printed circuits.

Characteristic Impedance the ratio of voltage or current in a propagating wave, i.e., the impedance that is offered to this wave at any point of the line. (In printed wiring its value depends on the width of the conductor, the distance from the conductor to ground plane(s), and the dielectric constant of the media between them.)

Chase a vise-like tool or instrument with adjustable, flexible draw bars for prestretching wire-mesh cloth prior to installing it in a screen printing frame. Also the frame used to hold the screen mesh for use on a screen printer.

Chip the unpackaged electronic semiconductor micro component, usually made of silicon. The chip is also known as an integrated circuit (IC) or die.

Chip-and-Wire a technology originally used in hybrids, where semiconductor chips are bonded face up and interconnected to the circuit by wire bonding.

Chip Carrier CC; a packaging system for electronic chips (IC's). The carrier provides protection and a practical means of connecting to circuitry.

Chip On Board COB; direct mounting of an electronic chip onto a circuit board without the usual hermetic package or chip carrier.

Circuit a combination of electrical and/or electronic components and an interconnecting network that accomplishes a desired function when activated with an electrical current.

Circuit Diagram conceptual diagram of electronic components and the relationship between them.

Circuit Element a basic constituent of a circuit, exclusive of interconnection.

Cladding a relatively thin layer of metal laminated to a core to form a base circuit board material.

Clamshell Press a screen printing press that is hinged at one end and opens like a clamshell. Can also refer to circuit board testers.

Clean Room a special manufacturing, assembly, or test area where filtered air and other cleaning processes are employed to minimize dirt.

Clinch mechanically securing an electronic device by bending over the leads that extend through a hole, prior to soldering or instead of soldering.

Clinched-wire through Connection a connection made by wire passed through a hole in a printed circuit board and subsequently formed, or clinched, in contact with the conductive pattern on each side of the board, and soldered.

Clip Terminal the point at which the hook-up wire is clipped against the connector post.

Closed Barrel Terminal a wire barrel terminal configuration available in strip form, or as loose pieces.

COB Chip On Board; see full term.

Coefficient of Linear Thermal Expansion the amount of expansion or increase in length occurring when a material is heated; expressed in inch/inch/°F. This term is important in flex circuit manufacturing and in electronic assemblies because a thermal mismatch will cause connection joints to break during temperature cycling.

Coefficient of Thermal Expansion change in dimensions as related to change in temperature. The CTE is important in circuitry, especially multilayers, where Z-axis expansion determines stress. Also critical for electronic assemblies because a mismatch can cause joint fracture over time. TCE values are often different for the X, Y, and Z axis of a material.

Cold Solder Joint a solder bonding connection made when the solder was already solidifying which causes an uneven solid structure which is typically unreliable.

Component Side that side of the printed board on which most of the components will be mounted.

Compression Connector an electrical connector that is crimped to make permanent. Usually a tube-like geometry.

Computer Aided Design CAD; hardware and software used to generate engineering, architectural and graphic arts drawings, diagrams, etc.

Computer Aided Engineering CAE; the use of moderately powerful computers to analyze and simulate designs to provide feedback for design refinement or verification.

Conductive Adhesive generally designates an electrically conductive bonding agent made by adding metals, such as silver, to epoxies and other adhesives.

Conductive Polymer a plastic material with intrinsic electrical conductivity or a dielectric one that is made conductive by adding particles of conductors such as carbon or metal.

Conductivity the ability of a material to conduct electric current. It is expressed in terms of the current per unit of applied voltage.

Conductor usually a metal, which is able to carry electrical current. In circuits, the etched metal or printed metal ink traces.

Conductor Width in printed circuits or cables, the edge to edge observable width of the metal conductors, viewed from above.

Conductor Spacing in printed circuits or cables, the distance between adjacent edges of conductors, usually measured in mils.

Conformal Coating an insulating protective coating, which conforms to the configuration of the object coated, applied to the completed board assembly.

Contact Bounce momentary rebounds occurring between two contact surfaces pushed together during switch actuation. Rating is the time between actuation and firm contact.

Contact Resistance the resistance in ohms between two objects in contact with each other.

Continuity an uninterrupted path for the flow of electrical current in a circuit.

Continuity Test an electrical test procedure where voltage is applied to those circuit conductors that should be interconnected and current flow is measured.

Controlled Collapse Chip Connection C4; an electronic packaging technology where tiny solder bumps are formed directly on Integrated Circuit (IC) chips, permitting them to be soldered directly to circuit boards. Also called ''flip chips.''

Coplanar an electrical impedance controlling configuration where a ground plane is placed adjacent to, and in the same plane as the signal line.

Copper Dust in printed circuits, an under-etch condition leaving very fine specks of copper on the circuit open area.

Copper Encirclement see Annular Ring.

Copper Mirror Test a method of determining the efficiency of solder flux. The flux is applied to a vacuum deposited copper mirror, which will be rapidly dissolved by good flux.

Copper Treatment a chemical applied to the surface of copper to inhibit tarnishing; Sealbrite or similar proprietary bath is used for copper printed circuits.

Corrosivity (circuitry) that characteristic of a conductor on a printed circuit which causes a resistance change greater than 10%, and/or causes it to show visible evidence of corrosion with discoloration when in physical touch with an (epoxy) adhesive, and is concurrently exposed to moderate humidity temperature effects, anytime after completed cure schedule.

Coupling the association of two or more circuits or systems in such a way that power may be transferred from one to another.

Cover Coat a dielectric coating applied to a circuit by screen printing. Used in place of a coverlay to provide protection and electrical insulation.

Cover Layer the outer or upper layer of insulating film applied over part or all of the conductor traces of a printed circuit.

Crosstalk interference caused by stray electrostatic (capacitative coupling) or electromagnetic (inductive) coupling of energy from one circuit to another.

Cure to change the physical properties of a material by chemical or physical process through the action of a catalyst such as heat, pressure, moisture or chemical reaction.

Curing Cycle for a thermosetting material, such as a resin, coating, or adhesive, it is the combination of total time-temperature profile to cause hardening by chemical cross-linking in an irreversible process.

Current Density the electrical current per unit area; amperes/sq.cm.

Current-Carrying Capacity the maximum current which can be carried continuously, under specified conditions, by a conductor without causing objectionable degradation of electrical or mechanical properties of the printed board.

Cylinder Press a modern type of screen printing press where flexible substrate is moved under the printing screen by a vacuum cylinder.

Datum Reference a defined point, line, or plane used to locate the pattern or layer for manufacturing, inspection, or for both purposes.

DC Direct Current; see full term.

Dead Rinse see Static Rinse.

Definition line sharpness or pattern reproduction accuracy in screen printing or photoimaging and any subsequent process like etching.

Deflux the removal of solder flux after soldering or roll tinning. Solvent dip or vapor condensation is commonly used. Aqueous systems are being adopted to reduce CFC and other solvent emissions.

Delaminate separation of foil from film or film from another film. A common defect with flexible circuits caused by heating too rapidly or without pre-drying the circuit.

Dendrite a metallic tree-like or nodular growth or plate out formed during electroplating at too high a current density. The phenomenon can also occur in a circuit if conditions are there for electromigration to occur. See treeing, silver migration.

Design Rule Checking DRC; refers to computerized check list in CAD (Computer Aided Design).

Designed for Testability DFT; circuits are laid out to facilitate electronic testing.

Deslug removal of material from holes and openings not completely punched out during hole fabrication when punch press methods are used.

Desoldering process of disassembling solder electronic parts to replace or repair.

DFT Designed for Testability; see full term.

Die Bond attachment of a die or chip to the hybrid substrate.

Die Bonding attaching a semiconductor chip to the substrate with adhesive, solder, or eutectic alloy.

Dielectric a material with very high resistance to the flow of electrical current. An insulator or nonconductor.

Dielectric Breakdown a complete failure of a dielectric material characterized by a disruptive electrical discharge through the material due to a sudden and large increase in voltage.

Dielectric Constant a material's ability to store a charge when used as a capacitor dielectric. It is the ratio of the charge stored with free space as the dielectric to that stored with the material in question as the dielectric.

Dielectric Layer a layer of dielectric material between two conductor plates, such as in a printed circuit.

Dielectric Loss the power dissipated by a dielectric as the friction of its molecules opposes the molecular motion produced by an alternating electric field.

Dielectric Properties the electrical properties of a material, such as insulation resistance, breakdown voltage, etc.

Dielectric Strength maximum voltage that an insulator (dielectric) can withstand before breaking down; volts per mil. Not a linear relationship although it varies with thickness.

Digitize to convert a drawing, diagram, or other entity to digital form using a computer input device, especially a digitizer pad.

Digitizer an input device to a computer which converts coordinate information to numerical data. Typically, the device is physically moved from point to point, and the movement is converted to data.

Dimensional Stability ability to retain shape under specified environmental conditions.

Din Plug a round, multipinned plug on either end of a computer device connection.

Direct Current DC; electrical current that is polarized and does not vary in sign. Battery output is DC, while house current is AC.

Direct Emulsion emulsion applied to a screen in a liquid form as contrasted to an emulsion that is transferred from a backing film of plastic (indirect).

Direct Emulsion Screen a screen whose emulsion is applied by precision coating directly onto the screen. This method produces the highest quality screen.

Double-Sided Substrate as the name implies, a substrate carrying active circuitry on both its topside and bottomside, electrically connected by means of metallized through-holes or edge metallization or both.

Dragout the solution that clings to material and racks when removed from solutions or processing baths.

Drag Soldering a component attachment process where a circuit board assembly is dragged across a molten, static pool of solder.

DRC Design Rule Checking; see full term.

Ductility that property which permits a material to deform without fracture.

Dumet Pin a round pin in packages, made of the core metal, usually Alloy 42, with copper cladding in the form of a thin sheath fused to the base metal made integral with it.

Dynamic Rinse a spray or flow rinse which is continually discharged to a drain.

Elastomeric Connectors a pliable strip of flexible material composed of alternating layers of electrical conductor and insulator intended for electrical interconnection of a device to a circuit, such as an electronic display to a circuit board. Sometimes called a zebra strip because of the striped appearance. Newer anisotropic products have a different look.

Electrical Insulation two conductors isolated from each other electrically by an insulating layer.

Electrical Noise (1) any unwanted disturbance within a dynamic electrical system. (e.g., undesired electromagnetic radiation in a transmission channel or device). (2) any unwanted electrical disturbance or spurious signal which modifies the transmitting, indicating, or recording of desired data.

Electrically Conductive Epoxy an organic polymer epoxy adhesive having metallic particles,such as gold, silver, palladium, etc. included as fillers in its formulation.

Electroless Plating the chemical deposition of a metal or alloy from an autocatalytic plating system without the use of electrical current.

Electrolytic Plating the deposition of an adherent metal or alloy onto a conductive substrate by applying current to the object and a counter electrode, both placed in a plating bath containing ions of the desired metal(s).

Electromagnetic Interference EMI; radiated energy from electrical and electronic devices which interferes with the operation of electronic circuitry. RF (radio frequency) signals are a common source of EMI.

Electromagnetic Welding a bonding process that uses electromagnetic energy, particularly in the radio frequency range, to activate adhesive materials. Typically, ferromagnetic particles which heat up when in the field, are incorporated into a thermoplastic adhesive causing it to melt and bond.

Electron Beam Bonding bonding two conductors by means of heating with a stream of electrons in a vacuum.

Electron Beam Patterning E-beam or EB; patterning of resist produces the required thin lines (as in VLSI) by evaporation from the heat supplied by the energy of a narrowly focused electron beam.

Electronic Design Interchange Format EDIF; a protocol for exchanging CAD and similar information between computers.

Electroplating deposition of an adherent metallic coating onto a conductive object placed into an electrolytic bath composed of a solution of the salt of the metal to be plated. Using the immersed object as the cathode and the other terminal as the anode (possibly of the same metal as the one used for plating), a DC current is passed through the solution affecting transfer of metal ions onto the cathodic surface.

Electrostatic refers to static electricity which occurs when an object is depleted or charged with electrons transferred by friction.

Electrostatic Charge an electric charge accumulated on an object, usually by friction between two objects or by transfer from another object.

Electrostatic Discharge ESD; the instantaneous transfer of charges accumulated on a nonconductor, along a conductor into ground caused either by direct contact or induced by an electrostatic field.

Elongation the degree of elasticity up to the point of rupture when the stress is applied to a material.

EMI Electromagnetic Interference; see full term.

Emulsion (screen printing) the light-sensitive material used to coat the mesh of a screen used to print. The emulsion is then imaged with light directed through a negative and then developed (unexposed emulsion removed) to create openings through which ink can pass.

Encapsulation sealing up or covering an element or circuit for mechanical and environmental protection. Liquid hardenable resins, such as epoxies or silicones, are typically used.

End Termination the metallization bands or metal end clips on the ends of discrete components or the metallization contacts on passive chip devices, provided for making electrical connections.

Environmental Test a test or series of tests used to determine the sum of external influences affecting the structural, mechanical, and functional integrity of any given package or assembly.

Epoxy Adhesive an organic polymer compound consisting of a thermosetting resin, a filler, a binder, a hardening agent, with a catalyst added, together with minor additives. Upon heating or the addition of catalyst, curing (polymerization) occurs producing a strong, hard bond.

ESD Electrostatic Discharge; see full term.

ESDS Electrostatic Discharge Sensitive.

Etch (1) a subtractive process whereby material, usually metal, is selectively removed by chemical reaction to produce precision parts, circuits, etc. In solid state electronics, microcircuits are etched chemically or by high energy beams, such as plasma (dry etching). (2) the action of an etchant, usually an acid solution or gas plasma, to remove unwanted portions of material on a surface by means of chemical or electrolytic action, producing an image, a delineated pattern, or an imprint.

Etch Back see Etchback.

Etch Factor the ratio of the depth of etch (conductor thickness) to the amount of lateral etch (undercut).

Etchant a solution used to remove, by chemical reaction, the unwanted portion of material from a printed board.

Etchback a process for the controlled removal of nonmetallic materials from sidewalls of holes to a specified depth. It is used to remove resin smear and to expose additional internal conductor surfaces.

Etching see Etch.

Eutectic the specific proportions of the components of an alloy having the lowest melting point. A eutectic alloy melts sharply. The alloy melting point is also called the eutectic or eutectic point.

Eutectic Alloy an alloy having the same temperature for melting and solidification, that is a sharp melting point which is also the lowest one for the particular components..

Eyeletting see Eyelet.

FA Failure Analysis.

Feed Through Device a wire-terminated component that is connected to a circuit board by feeding its wires, or leads, through one side of the circuit to the other side, followed by soldering.

Film an optional term for sheeting with a nominal thickness not greater than .010″.

Film Resistor (1) a deposited resistor (on a substrate), produced either from a thick film resistor composition by screen printing or from a thin film vacuum deposition. (2) an electrical resistor deposited by thin film or thick film techniques onto a circuit or a carrier.

Fineline a circuitry term used to indicate physical conductor geometries where width and spacings are less than .005″.

First Article a part or assembly manufactured prior to the start of production for the purpose of assuring that the manufacturer is capable of manufacturing a product which will meet the requirements.

First Level Interconnect interconnecting directly to an electronic chip.

Flexible a plastic film or composite with a modulus of elasticity not exceeding 10,000 PSI at 23°C, tested by Standard Flexure Stiffness method.

Flexible Printed Circuit an ordered pattern of electrical conductors on a flexible dielectric base film which becomes a dynamic part of an electronic circuit by means of its impedance characteristics. See Flexible Printed Wire.

Flexible Printed Wire FPW; an ordered pattern of electrical conductors on a flexible dielectric base film typically used as an interconnect cable. In the strictest sense, FPW only carries current and is not an active part of a circuit. See Flexible Printed Circuit.

Flexural Endurance a method used to determine the number of flexes to failure of flexible circuits.

Flip-Chip a monolithic IC structure designed to be directly connected to a circuit board usually by soldering. The C4 (Controlled Collapse Chip Connection) process, developed by IBM, forms solder bumps on the IC connection pads so that the device can be reflowed soldered to a circuit.

Floating Ground an electrical ground where signal and power ground are isolated.

Flood a moderately heavy coating or ink applied by screen printing a large area. Used to apply background color or special finishes. Also called Flood Coat.

Flood Bar the reciprocating horizontal bar on a screen printer which actively moves opposite to the direction of the squeegee with the purpose of returning the printing ink to its starting position in front of the squeegee.

Flux an agent that reduces metal oxide to fresh metal to facilitate soldering.

FMEA Failure Mode and Effects Analysis.

Foil the metal layer which is bonded to an insulator film to produce flexible circuit laminates.

Foot Print the conductor pads on a circuit board which are intended for attachment of a component; also called a land.

FPW Flexible Printed Wire; see full term.

Fusion Weld hot plate or bar welding where materials melt together.

Gang Bonding the process of making an array of connections simultaneously as in Tape Automated Bonding; see term.

Glass Transition Temperature T_g; the temperature at which an amorphous polymer (or the amorphous regions in a partially crystalline polymer) changes from a hard glassy and relatively brittle condition to a viscous or rubbery condition.

Glob 1) the material used to protectively encapsulate IC chips which have been attached directly to a circuit board (Chip on Board); usually an organic polymeric coating. 2) the process of globbing.

Green Growth growth or formation caused by the reaction of flux and copper oxide during roll tinning of Nomex circuits.

Ground in electronics it is the voltage reference point (0) in a circuit.

Ground Plane a conductive layer on a substrate or buried within a substrate that connects a number of points to one or more grounding electrodes.

Guide Pins pins, in the pin registration method, used to maintain proper alignment of substrate, artwork, masks, tools, and various other devices during processing.

H-Field an electromagnetic field where the magnetic component is stronger than the electric component.

Hard Solder solder that has a melting point above 800°F (425°C).

Hardener a chemical added to an adhesive or resin to promote curing or hardening by chemical reaction.

Hardness property of an ink or coating film to resist indentation or penetration by a hard pointed object. See Pencil Hardness.

Hardwired usually refers to a modem which sends its signals directly down the telephone line.

Heat Column the heating element in a eutectic die bonder or wire bonder used to bring the substrate up to the bonding temperature. A die is an electronic chip or IC which can be connected to the outside world by an interconnect process, such as die bonding.

Heat Distortion Point the temperature at which a standard test bar defects .010″ under a stated load.

Heat Flux the outward flow of heat from a heat source.

Heat Sink the supporting member to which electronic components, their substrate, or their package bottom are attached. This is usually a heat conductive metal with the ability to rapidly transmit heat from the generating source (component).

Heat Soak heating a circuit over a period of time to allow all parts of the package and circuit to stabilize at the same temperature.

Hot Spot a small area on a circuit that is unable to dissipate the generating heat and, therefore operates at an elevated temperature above the surrounding area; an observable bright spot caused by light sources; uneven diffusion.

Humidity Resistance ability to resist humidity without degradation; typical tests are 100 hrs at 100°F at 100% relative humidity or 85°C at 85% relative humidity for 1 to 7 days.

Hybrid Circuit a microcircuit consisting of elements which are a combination of the film circuit type and the semiconductor circuit type, or a combination of one or both of these types that may include discrete add-on components.

I/O   Input/Output of computers and related equipment.

IBP   Integrated Battery Pack; see full term.

IC   Integrated Circuit; see full term.

ILB   Inner Lead Bond(er); see full term.

ILO   Injection-Locked Oscillator; an electronic device.

Imbedded Layer   a conductor layer having been deposited between insulated layers.

Immersion Plate   chemical deposition of metal over base metal achieved by the partial displacement of base metal; a form of electroless plating.

Impedance   the total opposition to current flow offered by a circuit or an electronic device and composed of resistance, inductance, and capacitance.

Impedance Control   the design technique of suppressing unintentional radiation by providing matched impedance conduction paths for electronic signals.

Inactive Flux   flux that becomes nonconductive after being subjected to the soldering temperature.

Inclusions   foreign particles in the conductive layer or base film layer.

Index   the processing or punching of registration holes in web or sheet prior to imaging or other processes requiring registration.

Indexing   see Index.

Indirect Emulsion   screen emulsion that is transferred to the printing screen surface from a plastic carrier or backing material. The emulsion is then exposed to light through a negative and developed to form a printing stencil.

Infrared   the band of electromagnetic wavelengths lying between the extreme of the visible (0.75 nm) and the shortest microwave (1000 nm). Warm bodies emit the radiation, bodies which absorb the radiation are warmed.

Initial Graphics Exchange Specification   IGES; an ANSI accepted specification for interchanging computerized drawings (CAD). It is oriented toward display and 2 dimensional drawings.

Initial Sample Inspection Report   ISIR; originated by Ford as a qualification step. Now used to qualify lots through early production.

Ink   a liquid, typically highly viscous material, which can be applied selectively to a substrate by a printing process. Ink consists of binder, either resin or inorganic, and usually a solid filler which imparts aesthetic qualities (graphics) or electrical characteristics (thick film ink) as well as rheological properties.

Inner Lead Bond   ILB; refers to Tape Automated Bonding bonds made to an integrated circuit.

Input/Output   I/O; refers to the input output porting in computers and other electronic systems. (2) communication ports on computers and other electronic systems.

Insulation Resistance   resistance to current leakage offered by insulation material to a DC potential.

Insulators   a class of materials with high resistivity. Materials that do not conduct electricity. Materials with resistivity values of over 10^5 ohms are generally classified as insulators.

Integrated Battery Pack   IBP; an electronic chip with a self-contained battery for backing up memory.

Integrated Circuit   IC; a microcircuit (monolithic) consisting of interconnected elements inseparably associated and formed in situ or within a single substrate (usually silicon) to perform an electronic circuit function.

Interconnection the conductive path required to achieve connection from a circuit element to the rest of the circuit.

Interface (1) the boundary between dissimilar materials, such as between a film and substrate or between two films.

Interfacial Bond an electrical connection between the conductors on the two faces of a substrate.

Intermetallic Bond the ohmic contact made when two metal conductors are welded or fused together.

Intermetallic Compound a compound of two or more metals that has a characteristic crystal structure that may have a definite composition corresponding to a solid solution, often refractory. Intermetallic bonds formed in soldering, usually brittle.

International Standards Organization ISO; major world standards group. ANSI (American National Standards Institute) is the United States' member body.

Ion Migration the movement of free ions within a material or across the boundary between two materials under the influence of an applied electric field.

Ion Milling bombardment of a surface by high energy ions in order to remove material very precisely.

ISIR Initial Sample Inspection Report; see full term.

ISO International Standards Organization; see full term.

ITO Indium Tin Oxide; a partially oxidized alloy of tin and indium metal, applied to a substrate in a very thin, often optically clear form. The coating is electrically conductive.

JC JEDEC Committee.

JEDEC Joint Electron Devices Engineering Council; see full term.

Joint Electron Devices Engineering Council JEDEC; a standards setting association for electronic packaging.

Jumper a direct electrical connection between two points on a film circuit. Jumpers are usually portions of bare or insulated wire mounted on the component side of the substrate.

Junction (1) a contact made between dissimilar metals (2) a connection made between two or more electrical conductors (3) a connection made between two conductors and conductive adhesive.

K a symbol for dielectric constant.

K Factor refers to thermal conductivity, the ability of a substance to conduct heat through its mass.

Kerf the slit of channel cut in a resistor during trimming by laser beam or abrasive jet.

Kovar a metal alloy with very low thermal expansion approaching that of glass and ceramics; 17% Co, 53% Fe, 29% Ni, and trace metals.

Ladder Network a series of film resistors with values from the highest to the lowest resistor reduced in known ratios.

Laminate (verb) the process of bonding two or more layers of material together, typically using adhesive under heat and pressure: (noun) a layered sandwich of sheets of substances bonded together under heat and pressure to form a single structure.

Land   a conductive area on a printed circuit board for the attachment of electronic components.

Land Pattern   a combination of lands (component mounting areas) intended for the attachment of a particular component.

Laser Bonding   effecting a metal-to-metal bond of two conductors by welding the two materials together using a laser beam for a heat source.

Layout   the positioning of the conductors and/or resistors on artwork prior to photoreduction of the layout to obtain a working negative or positive used in screen preparation.

LCC   Leadless Chip carrier; see full term.

LCCC   Leadless Ceramic Chip Carrier; see full term.

LCD   Liquid Crystal Display; see full term.

Leaching   in soldering, the dissolving (alloying) of the material to be soldered into the molten solder. Generally undesirable.

Lead   (1) a conductive path which is usually self-supporting. Electronic packaged devices are usually provided with leads for attachment to circuits and are referred to as leaded devices; (2) a wire with or without terminals that connects two points in a circuit.

Lead Frames   the metallic portion of the device package that completes the electrical connection path from the die or dice and from the ancillary hybrid circuit elements to the outside world.

Leadless Ceramic Chip Carrier   LCCC; a surface mount package, having no lead edge terminations and constructed of ceramic material.

Leadless Chip Carrier   LCC; a surface mount electronic component package having metal terminations on all four sides instead of leads.

Leadless Device   an electronic package used to contain an integrated circuit (IC) or other electronic device, that has connector terminations attached directly to the housing instead of wire leads.

Leakage Current   an undesirable small stray current which flows through or across an insulator between two or more electrodes, or across a back-biased junction.

Leveling   a term describing the settling or smoothing out of the screen mesh marks in thick films that takes place after a pattern is screen printed. May also be applied to coatings.

Light-Emitting Diode   LED; a solid state device capable of emitting visible or infrared light when current is applied. Typically, the bright red, amber or green indicator lights on digital displays.

Line Definition   a measure of the sharpness or cleanness of screen printed lines. The precision of line width is determined by twice the line edge definition/line width. A typical precision of 4% exists when the line edge definition/line width is 2%.

Liquid Crystal Display   LCD; common alphanumeric/graphic display used in portable computers, watches etc. Very low power consumption. Controls rather than emits light.

Liquidus the line on a phase diagram above which the system has molten components. The temperature at which melting starts.

Live Rinse see Dynamic Rinse.

Loss Tangents the decimal ratio of the irrecoverable to the recoverable part of the electrical energy introduced into an insulating material by the establishment of an electric field in the material.

Low Loss Substrate a circuit substrate with high radio-frequency resistance and hence, slight absorption of energy when used in a microwave integrated circuit.

Machine Direction the direction that the web travels on a machine, such as a coater or laminator. Opposite of transverse direction.

Machine Vision MV; optical system that compares and makes a quality decision.

Mask (1) the photographic negative that serves as the master for making thick-film screens and thin-film patterns (2) the ultra-precision exposure masks used to fabricate integrated electronic circuits.

MCM Multi-Chip Module; see full term.

Mean Time Between Failures MTBF; arithmetic mean average time (hours), which can be expected between device failure.

Mean Time to Failure MTTF; (1) applicable to individual parts or devices in reliability testing (2) statistically averaged time before a component is expected to fail.

MEFB Metal Electrode Face Bonding; see full term.

MELF Metal Electrode Face; see full term.

Melinex a trade name for polyester film. See Mylar, Polyester.

Mesh See Mesh Size.

Mesh Size (1) the number of openings per inch in a printing screen. A 200 mesh screen has 200 openings per linear inch, 40,000 openings per inch. See Screen Printing. (2) the number of wires per inch in sieving screens.

Metal Electrode Face MELF; a surface mountable electronic package that is cylindrical in shape and is soldered at the end caps. See also, Surface Mount Technology (SMT).

Metal Electrode Face Bonding MEFB; same as Metal Electrode Face (MELF), which is the more common term because of pronouncability.

Metal On Elastomer MOE; an electrical interconnect strip made up of alternating slices of metal conductor and dielectric rubber. An electrical connection is made by sandwiching the strip between a circuit board and a device; also called Zebra strip.

Metallization (1) (electronic) a film pattern (single or multilayer) of conductive material deposited on a substrate to interconnect electronic components, or the metal film on the bonding area of a substrate which becomes a part of the bond and performs both an electrical and a mechanical function (2) (general) vacuum deposited thin metal film.

MFD Microelectronic Functional Device.

Microbond a bond of a small wire such as 0.001 inch in diameter gold, to a conductor or to a chip device.

Microcircuit a small circuit having a relatively high equivalent circuit element density. This excludes monolithic integrated circuits, hybrids, printed wiring boards, circuit card assemblies, and modules composed exclusively of discrete electronic parts.

Microcomponents small discrete components, such as chip transistors, resistors, and capacitors.

Microelectronics that area of electronic technology associated with or applied to the production of electronic systems from extremely small electronic parts or elements.

Micron μ; a metric unit of length equal to a micrometer or one millionth of a meter; 10^{-6} meters. 25 microns = .001 inch.

Microstrip (1) a microwave transmission component usually on a ceramic substrate (2) an electrical impedance controlling configuration where ground plane is placed above and below a signal line.

Migration an undesirable phenomenon whereby metal ions, notably silver, are transmitted through another metal, or across an insulated surface, in the presence of moisture and an electrical potential.

Mil (1) 1/1000 of an inch (.001″); English units commonly used in circuitry (2) a prefix on military specifications, hence mil can refer to military specs or components.

MMC Metal Matrix Composite; see full term.

Modular Automatic Test Equipment MATE; circuit board and electronic assembly computerized test equipment which can be mated together as the test need arises.

MOE Metal On Elastomer; see full term.

Moisture Stability a measure of the dimensional and material stability of a circuit under high humidity conditions.

Molded Printed Circuit Board MPCB; an electronic circuit whose nonconductive base is made from injection molded plastic. MPCB's may be planar or three-dimensional.

Mother Board a circuit board possessing ''female'' plug-in slots used to interconnect smaller circuit boards called ''daughter boards.'' The main circuit board.

MPCB Molded Printed Circuit Board; see full term.

MSL Micro Strip Line; a planer asymmetric transmission system for high frequency electronic signals.

MTBF Mean Time Between Failures; see full term.

MTTF Mean Time To Failure; see full term.

Multichip Module MCM; a microelectronic circuit consisting solely of active devices and passive chips which are separately attached to the major substrate and interconnected to form the circuit.

Multilayer Circuit a composite circuit consisting of alternate layers of conductive circuitry and insulating materials (ceramic or dielectric compositions) bonded together with the conductive layers interconnected as required.

Multilayer Substrate substrates that have buried conductors, several planes of conductors, allowing complex circuitry to be layed out.

Multiple Circuit Layout layout of an array of identical circuits on a substrate. Also called multi-up.

Multi-Up See Multiple Circuit Layout.

Mylar DuPont's trade name for polyester film; type D is very clear and used for graphics. Type EL, electrical grade, used for membrane switches and flexible circuitry. See Polyester.

NC (1) in machinery, numerical control: microprocessor programmable equipment. (2) in switches, normally closed.

NDT Non-Destructive Test.

Negative N; an artwork, artwork master, or production master in which the intended conductive pattern is transparent to light, and the areas to be free from conductive material are opaque.

Negative Image reverse artwork where originally clear areas are now opaque.

Negative Temperature Coefficient the device changes its value in the negative direction with increased temperature.

N-Key Rollover the capability of a keyboard system to decipher the value of a key when another is pressed simultaneously.

nm nanometer; see full term.

Noble Metal unreactive, precious metallic elements including gold, silver, platinum, rhodium, and palladium.

Noble Metal Paste paste materials composed partially of noble metals, such as gold or ruthenium.

Nodule (1) inclusion of foreign particles in or onto screen printed material, producing a defective image (2) the plating over foreign particles resulting in surface protrusions.

Nominal Resistance Value the specified resistance value of the resistor at its rated load.

Non-Recurring Engineering NRE; the engineering, and subsequent cost, incurred at the inception of a program but not repeated.

Non-Rigid see Flexible.

NRE Non-Recurring Engineering; see full term.

Nugget the region of recrystallized material at a bond interface which usually accompanies the melting of the materials at the interface.

Occluded Contaminants contaminants that have been absorbed by a material.

OCR Optical Character Recognition; see full term.

Off Bond bond that has some portion of the bond area extending off the bonding pad.

Off Contact Printing normal screen printing where the screen is set to a certain height or off contact distance above the substrate to be printed. Movement of the squeegee across the screen brings the mesh momentarily in contact.

Off Contact Screener a screen printer that uses off contact printing of patterns onto substrates as compared to a direct contact printer.

Ohm a unit of electrical resistance. A 60 ft. length of 22 gage copper wire has a resistance of 1 ohm. Circuit resistance produces a drop in voltage. Resistance = voltage/current.

Ohm-Cm a unit of material volume resistivity.

Ohm's Law a mathematical statement which defines how electrical current, voltage and resistance are related. Current is directly proportional to voltage but inversely proportional to resistance: I(current) = E(voltage)/R(resistance)

Ohm/Square/Mil a measure of electrical volume resistivity often used for PTF inks. A square is a dimensionless unit of equal length and width. Square/mil is a volume unit. Mil = .001″.

Ohmic Contact a contact that has linear voltage current characteristics through out its entire operating range.

Ohm Meter an electrical test device used to measure electrical resistance, the ohm. See Ohm, Resistance.

Ohms/Square a unit of material surface resistivity where length and width (a square) are equal.

OLB Outer Lead Bond; see full term.

Open an interruption in an electrical pathway such as in a circuit board so that current flow is incomplete or halted.

Optical Character Recognition OCR; page reader for computer which digitizes the alphanumeric information.

Outer Lead Bond OLB; a Tape Automated Bonding (TAB) interconnect circuit, consists of inner leads that are connected to an electronic chip, and outer leads, which are connected to a circuit board. The latter bonds are referred to as OLB's. See Tape Automated Bonding.

Outgas the release of gas from a material over a period of time or during heating.

Overcoat a thin film of insulating material, either plastic or inorganic (e.g., glass or silicon nitride) applied over integral circuit elements for the purposes of mechanical protection and prevention of contamination.

Overhang the sum of the plating outgrowth and the etching undercut. Typically, solder plate on copper results in a solder overhang.

Overlap the contact area between a film resistor and a film conductor.

Overlay the graphically decorated upper layer of a graphic display/control panel. Also called a graphic overlay or Face Plate. Incorrectly called a label.

Overtravel the excess downward distance a squeegee blade would push the screen if the substrate were not in position.

Oxidizing Atmosphere an air, or other oxygen-containing atmosphere in a firing furnace which oxidizes the resistor materials while they are in the molten state, thereby increasing their resistance.

Package the container for an electronic component(s) with terminals to provide electrical access to the inside of the container. In addition, the container usually provides hermetic and environmental protection for, and a particular form factor to, the assembly of electronic components.

Packaging Density quantity of functions (components, interconnection devices, mechanical devices) per unit volume, usually expressed in qualitative terms, such as high, medium or low.

Pad (1) a metallized area on the surface of an active substrate as an integral portion of the conductive interconnection pattern to which bonds or test probes may be applied (2) generally rectangular conductor shapes where components or wires are attached, usually by soldering.

Pads Only the process where only exposed pads on a circuit are plated.

Panel Plate the plating of the entire surface of a panel or web, including holes (plated through holes).

Paralene a polymer series of coatings, applied in a vacuum chamber to circuits to provide efficient protection even in a very thin form.

Parallel Gap Solder passing a high current through a high-resistance gap between electrodes to re-melt solder thereby forming an electrical connection.

Parallel Gap Weld   passing a high current through a high-resistance gap between two electrodes that are applying force to two conductors, thereby heating the workpieces to the welding temperature and effecting a welded connection.

Parasite Losses   losses in a circuit often caused by the unintentional creation of capacitor elements in a film circuit by conductor crossovers.

Partial Lift   a bonded lead partially removed from the bonded area.

Passive Components   elements (or components) such as resistors, capacitors, and inductors which do not change their basic character when an electrical signal is applied. Transistors are the most common active components.

Passive Device   see Passive Components.

Passive Network   a circuit network of passive elements such as film resistors that are interconnected by conductors.

Paste   synonymous with "composition" and "ink" when relating to screenable thick-film, Polymer Thick Film, or solder materials.

Paste, Soldering   finely divided particles of solder suspended in a flux paste. Used for screening application onto a film circuit and reflowed to form connections to chip components.

Pattern   the outline of a collection of circuit conductors that defines a circuit board or network.

Pattern Plate   the selective plating of a conductive pattern/on a circuit or laminate.

Pattern Recognition   a computerized system capable of analyzing video input and distinguishing shapes.

PCA   Printed Circuit Assembly.

PCB   Printed Circuit Board.

Peel Bond   similar to lift-off of the bond with the idea that the separation of the lead from the bonding surface proceeds along the interface of the metallization and substrate insulation rather than the bond-metal surface.

Peel Strength   a measure of adhesion between a conductor and the substrate. The test is performed by pulling or peeling the conductor off the substrate and observing the force required. Units are oz/mil or lb/in. of conductor width (2) the force/unit width required to separate two bonded layers; typically metal foil from base laminate at an angle of 90° or 180°.

Pencil Hardness   technique for measuring hardness. Pencil is sharpened to chisel point and pushed at a 45° angle to surface of coating. The higher the letter/number the pencil that the coating resists, the harder the coating.

PGA   (1) Pin Grid Array; see full term (2) Programmable Gain Amplifier; see full term.

Phosphor Bronze   a strong, durable alloy used particularly for springs.

Photo Etching   the process of forming a circuit pattern in metal film by light hardening a photosensitive plastic material through a photo negative of the circuit etching away the unprotected metal.

Photomask   a glass- or film-based material, used as an image carrier, capable of maintaining high dimensional stability and accurate tolerances.

Photoresist   a photosensitive plastic coating material which when exposed to UV light becomes hardened and is resistant to etching solution. Typical use is as a mask in photochemical etching of thin films.

Pierce process of punching holes for components into a circuit web or sheet.

Pierce and Blank the simultaneous punching of component holes and cutting out of circuits using a punch press.

Pierce Die a tooling die producing openings or holes in a substrate by punching out ''slugs.''

Pigtail a wire bond term that describes the excess wire that remains at a bond site beyond the bond. Excess pigtail refers to remnant wire in excess of three wire diameters.

Pin Grid Array PGA; a package or interconnect scheme, featuring a multiplicity of plug-in type electrical terminals arranged in a prescribed matrix format or array. Used with high I/O-count devices.

Pin Staking the procedure of mechanically inserting pins into a circuit.

Pinheads small pinhead-like protrusions of plating that result from pinholes in the plating resist.

Pinhole small holes occurring as imperfections which penetrate entirely through film elements, such as metallization films or dielectric films.

Pinout output sequence of electronic devices, circuits, and membrane switches.

Pit (1) a depression produced in a metal surface produced by uneven electrodeposition or by corrosion (2) depression in conductive layer that doesn't penetrate entirely through (3) depressions produced in metal or ceramic surfaces by non-uniform deposition.

Pitch the nominal distance from center-to-center of adjacent conductors. (Where conductors are of equal size and spacing is uniform, the pitch is usually measured from the reference edge of a conductor to the referenced edge of the adjacent conductor.)

Plasma a partially ionized gas in which equal numbers of positively and negatively charged particles coexist with neutral (uncharged) particles, usually in a vacuum as the result of the application of energy (such as radio frequency; RF). Can be used to clean and activate surfaces for bonding.

Plastic a polymeric material, either organic (e.g., epoxy) or silicone used for conformal coating, encapsulation, or overcoating.

Plastic Device a device wherein the package, or the encapsulant material for the semiconductor die, is plastic. Such materials as epoxies, phenolics, silicones, etc., are included.

Plastic Encapsulation environmental protection of a completed circuit by embedding it in a plastic such as epoxy or silicone.

Plastic Laser Machining a laser is used to fracture molecules in the polymer chain, not to melt plastic as in laser drilling. Ultraviolet range excimer lasers are used.

Plastic Leaded Chip Carrier PLCC; (1) a package for an electronic chip of rectangular shape which has metal leads protruding through the plastic case which are connected to a circuit board (2) an electronic package of the SMD type, made of plastic and having terminations on all four sides. There is some confusion associated with this acronym because L can stand for ''leaded'' or ''leadless.''

Plastic Quad Flat Pack PQFP; a new Surface Mounted Device with higher density and lower profile.

Plastic Shell a thin plastic cup or box used to enclose an electronic circuit for environmental protection or used as a means to confine the plastic encapsulant used to imbed the circuit.

Plastisol a suspension of a resin in a plasticizer which can be used as a coating, adhesive or casting material. Heating hardens the mixture.

Plated Through Hole PTH; a hole in a circuit used to connect a double-sided circuit together by creating a connection by plating metal on the walls of the hole.

Plating Up the process consisting of the electrochemical deposition of a conductive material on the base material (surface holes, etc.) after the base material has been made conductive.

PLCC Plastic Leaded Chip Carrier; see full term.

Plotting the practice of mechanically converting X-Y positional information into a visual pattern, such as artwork.

Plug-In-Package an electronic package with leads strong enough and arranged on a surface so that the package can be plugged into a test or mounting socket and removed for replacement as desired without destruction.

PN Junction the junction at which a P-type and an N-type semiconductor are joined to produce a solid state electronic device.

PNP a solid state transistor device produced by joining P-type, N-type, and P-type semiconductor materials together in that order.

Point-To-Point Wiring an interconnecting technique wherein the connections between components are made by wires routed between these points.

Polycrystalline a material is polycrystalline in nature if it is made of many small crystals. Alumina ceramics are polycrystalline, whereas glass substrates are not.

Polyester a resin formed by polymerizing a difunctional acid with a dihydroxy alcohol, producing ester links; Mylar and Melinex are trade names.

Polyimide a class of polymer or resin containing nitrogen-based imide groups used for high temperature substrates and adhesives.

Polymer Bonding the simultaneous electrical and mechanical connecting of components to a circuit with polymers, such as conductive epoxy.

Polymer Thick Film selectively depositing, by screen printing or other additive processes, liquid material which is dried or cured to impart desired electrical properties: e.g.; silver conductive ink. PTF is usually thicker than .0005″ and "thick" distinguishes it from "thin" vacuum deposition.

Poly-Solder a conductive adhesive for bonding surface mounted devices to circuits that are electrically stable under environmental conditions.

Positive an artwork, artwork master, or production master in which the intended conductive pattern is opaque to light, and the areas intended to be free from conductive material are transparent.

Positive Image the true picture of a circuit pattern as opposed to the negative image or reversed image.

Positive Temperature Coefficient the changing of a value in the positive direction with increasing temperature.

Post Curing heat aging of a film circuit after firing to stabilize the resistor values through stress relieving.

Post Stress Electrical the application of an electrical load to a film circuit to stress the resistors and evaluate the resulting change in values.

Pot Life useful working life of a catalyzed adhesive, coating, or other such material.

Potting encapsulating of a circuit in plastic, usually a thermoset liquid.

Power Density the amount of power dissipated from a film resistor through the substrate measured in W/in^2.

Power Dissipation the dispersion of the heat generated from a film circuit when a current flows through it.

PQFP Plastic Quad Flat Pack; see full term.

Prefired conductors fired in advance of the screening of resistors on a substrate.

Preform as an aid in soldering, small circles or squares of the solder are punched out of thin sheets. These preforms are placed on the spot to be soldered or bonded, prior to the placing of the object to be attached.

Prepreg sheet material (e.g., glass fabric) impregnated with a resin cured to an intermediate stage (B-stage resin).

Pressure Sensitive Adhesive PSA; a highly tacky dry adhesive film requiring only pressure, not heat or time, to bond two substrates together.

Print and Fire a term sometimes used to indicate steps in the thick-film process wherein the ink is printed on a substrate and is heated at high temperature (fired).

Print Laydown screen printing of the film circuit pattern onto a substrate.

Printed Board the general term for completely processed printed circuit or printed wiring configurations. It includes rigid or flexible, single, double and multilayer boards.

Printed Circuit (1) a generic term used to describe a printed board produced by any of a number of techniques including non-printing processes like photo resist etching (2) a conductive pattern comprised of printed components, printed wiring, or a combination thereof, all formed in a predetermined design and intended to be attached to a common base.

Printed Component a part, such as an inductor, resistor, capacitor, or transmission line, which is formed as part of the conductive pattern of the printed board.

Printed Through Hole an electrical connection made from one side of a circuit board to the other by means of conductive ink which is printed into and through a hole to the opposite side. Also called Screened Through Hole, the name derived from screen printing.

Printed Wire Board PWB; the common printed circuit board.

Printing Parameters the conditions that effect the screening operation, such as off-contact spacing, speed and pressure of squeegee, etc.

Private Automatic Branch Exchange PABX; internal business phone exchange.

Probe a pointed conductor used in making electrical contact to a circuit pad for testing.

Probotics the technology of designing, fabricating, and interfacing probes to a mechanical device or system operated manually or automatically for such purposes as microcircuit testing or laser trimming.

Profile a graph of time vs. temperature, or of position in a continuous thick film furnace vs. temperature. Term can apply to reflow soldering or any other oven process.

Profilometer   a contact profiling instrument which measures surface roughness or topology.

PSA   Pressure Sensitive Adhesive; see full term.

PSG   Phospho-Silicate Glass.

PTH   Printed Through Hole; see full term.

Pull Strength   the values of the pressure achieved in a test where a pulling stress is applied to determine breaking strength of a lead or bond.

Pull Test   a test for bond strength of a lead, interconnecting wire, or a conductor.

Pulse Soldering   soldering a connection by melting the solder in the joint area by pulsing current through a high-resistance point applied to the joint area and the solder.

Push-Off Strength   the amount of force required to dislodge a chip device from its mounting pad by application of force to one side of the device, parallel to the mounting surface.

PWB   Printed Wire Board; see full term.

Pyrolyzed   a material that has gained its final form by the action of heat is said to be pyrolized. Broken down by heat but not burned (oxidized).

Radiated Emission   radiation of electromagnetic fields into space.

Radio Frequency Tag   an electronic tag capable of receiving, storing, and transmitting digital information by means of and in response to RF energy.

Random Failures   circuit failures which occur randomly with the overall failure rate for the sample population being nearly constant.

RC Network   a network composed only of resistors and capacitors.

Real Estate   the surface area of an integrated circuit or of a substrate. The surface area required for an electronic component or element.

Red Stains   a condition of accelerated adhesive breakdown during platen pressing of solder-plated circuits that were insufficiently rinsed after a high chloride process.

Reflow or Reflowing   the melting or re-melting of metal or polymers by heating so as to smooth out the material or to bond.

Reflow Soldering   a method of soldering involving application of solder prior to the actual joining. To solder, the parts are joined and heated, causing the solder to remelt or reflow.

Registration   (1) the alignment of a circuit or graphic pattern to the geometry of the substrate and/or the alignment of subsequent images to the first one (2) the degree of placement conformity; actual vs. intended (3) exactness of fit or closeness of tolerance color-to-color or color-to-border.

Registration Marks   the marks used for aligning successive processing masks by visual or optical methods. Typically, cross-hair targets.

Risers   the conductive paths, or conduits, that run vertically from one layer to another, in multilayer circuits.

Resist   (1) a coating, ink or dry film used to mask selected areas from the action of etchant, plating, solder, and various vapor or liquid processes (2) a protective coating that will keep another material from attaching or coating something, as in solder resist, plating resist, or photoresist.

Resistance   R; the tendency of any material to resist the flow of electrical current with the resulting conversion to heat. See Ohm, Resistivity.

Resistance Soldering a method of soldering in which a current is passed through and heats the soldering area by contact with one or more electrodes.

Resistance Weld the joining of two conductors by heat and pressure with the heat generated by passing a high current through the two mechanically joined materials.

Resistive Temperature Detector RTD; a general term for electrical devices which sense temperature by undergoing corresponding changes in electrical resistance; a thermister.

Resistivity a proportionality factor characteristic of different substances equal to the resistance that a centimeter cube of the substance offers to the passage of electricity, the current being $R = \rho L/A$, where R is the resistance of a uniform conductor, L its length, A its cross-sectional areas, and ρ its resistivity. Resistivity is usually expressed in ohm-centimeters.

Resistor an electrical device that provides a means of limiting current flow and dropping voltage.

Resistor Array a chip containing a number of resistors and designed to mount onto circuits.

Resistor Drift the change in resistance of a resistor through aging and usually rated as percent change or parts per million per 1000 hr.

Resistor Geometry the film resistor outline. See Aspect Ratio.

Resistor Overlap the contact area between a film resistor and a film conductor.

Resolution (1) the degree of fineness or detail of an imaging process, such as screen printing or photoexposure (2) the degree of detail in a video display.

Reverse Current (Anodic) Cleaning electrolytic cleaning in which a current is passed between electrodes through the solution with the part being the anode.

Reverse Image the resist pattern on a printed board used to allow for the exposure of conductive areas for subsequent plating.

Rework all work performed, other than testing, in a hybrid microcircuit (after initial circuit fabrication) on parts with known deficiencies so as to cause such parts or the entire microcircuit to comply fully with documented requirements.

RF Radio Frequency.

RFI Radio Frequency Interference.

Rheology the science dealing with deformation and flow of matter.

Ribbon Interconnect a flat narrow ribbon of metal such as nickel, aluminum, or gold used to interconnect circuit elements or to connect the element to the outer pins.

Ribbon Wire metal in the form of a very flexible flat thread or slender rod or bar tending to have a rectangular cross-section as opposed to a round cross-section.

Risers the conductive paths that run vertically from one level of conductors to another in a multilayer substrate or screen printed film circuit.

Risetime the time required for the output voltage of an electronic device to rise from 10% to 90% of the final high voltage level.

Roll Over Circuit an electronic device that provides temporary storage of keyboard input signals to prevent error when two or more keys are pressed in rapid succession.

Rosin Flux a flux having a rosin base which becomes interactive after being subjected to the soldering temperature.

Scale an oxide coating on metal such as copper circuitry, heavier than tarnish.

Scaling peeling of a film conductor or film resistor from a substrate, indicating poor adhesion.

Scallop Marks a screen printing defect which is characterized by a print having jagged edges. This condition is a result of incorrect dynamic printing pressure or too thin an emulsion. The scallop pattern is produced by the screen mesh and it can be used to determine the mesh count of the screen that was used.

Scanning Electron Microscope SEM; a powerful imaging microscope which uses an electron beam instead of light to produce very high magnification. A highly versatile instrument for diagnostic microscopy when applying surface microanalytical techniques.

Scavenging in vacuum metallizing, placement of a gettering metal into the vacuum chamber to scavenge undesirable materials, particularly gasses. Also see Leaching.

Schematic a diagram of a functional electronic circuit consisting of symbols for all active and passive elements with their interconnecting matrix that forms the circuit.

Screen a network of metal or fabric mesh, mounted tautly on a frame, and upon which a stencil of patterns and configurations are generated, usually by photographic means. The screen is used as the ''printing plate'' in the screen printing process.

Screen Deposition the laydown, or printing, of an ink pattern on a substrate using the screen printing technique.

Screen Frame in screen printing, the metal frame that holds the screen fabric taut to permit it to be used for printing. Also called a chase.

Screen Printing method of transferring an ink, paste or other liquid onto a substrate by pressing the liquid through a stencil pattern screened by pulling a squeegee across the surface of the screen. Also called screening and silk screen printing (obsolete).

Selective Etch restricting the etching action on a pattern by the use of selective chemicals which attack only one of the exposed materials.

SEM Scanning Electron Microscope; see full term.

Semi-Additive a circuit process where a thin layer of conductor (flashed or metallized on) is etched (subtractive) and then plated up (additive).

Semiconductor a material having electrical characteristics falling between conductors and insulators exemplified by carbon, silicon and germanium. Electronic amplification and control devices, such as transistors, diodes, and integrated circuits can be fabricated using specially prepared semiconductors also known as solid state devices.

Serpentine Cut a trim cut in a film resistor that follows a serpentine or wiggly pattern to effectively increase the resistor length and increase resistance.

Shadows a very thin layer of screened-on resist, extending over the edge of the conductors; due to thin ink, bleeding, or incorrect screen set up.

Shear Rate the relative rate of flow or movement (of viscous fluids).

Shear Strength the limiting stress of a material determined by measuring a strain resulting from applied forces that cause or tend to cause contiguous parts of a body to slide relative to each other in a direction parallel to their plane of contact; the value of the force achieved when shearing stress is applied to the bond (normally parallel to the substrate) to determine the breaking load.

Sheet Resistance the electrical resistance of a thin sheet of a material with uniform thickness as measured across opposite sides of a unit square pattern.

Shelf Life the maximum length of time, usually measured in months, between the date of shipment of a material to a customer and the date by which it should be used for best results.

Shield a screen or other housing (usually conductive) placed around devices or circuits to reduce the effect of electric or magnetic fields on them.

Shielding a physical barrier, usually electrically conductive, designed to reduce the interaction of electric or magnetic fields upon devices, circuits, or portions of circuits.

Short Term Overload a circuit that has been overloaded with current or voltage for a period too short to cause breakdown of the insulation.

Signal Plane a conductor layer intended to carry signals, rather than serve as a ground or other fixed voltage function.

Signal an electrical impulse of a predetermined voltage, current, polarity, and pulse width.

Silicone a polymeric material composed of recurring silicon and oxygen groups. Silicones can be produced as fluids, greases or rubbers.

Silk Screen Printing see Screen Printing.

Silver Migration formation of short circuiting pathways of silver between silver conductors exposed to moisture and carrying DC current; silver dendrites.

Single In-Line Package SIP; an electronic package or container with connection pins in a single row and protruding from one edge of the package. SIP's often consist of circuits containing a number of components.

Sinking shorting of one conductor to another on multilayer screen printed circuits because of a downward movement of the top conductor through the molten crossover glass.

SIP Single In-Line Package; see full term.

Skin Effect the increase in resistance of a conductor at microwave frequencies because of the tendency for current to concentrate at the conductor surface.

Slivers metal overhang, as the result of etching, that breaks off. Usually refers to solder plate.

Slump a spreading of printed thick-film composition after screen printing but before drying. Too much slumping results in loss of definition. Also refers to solder paste or conductive adhesive spreading after stenciling or printing.

Small-Outline Integrated Circuit SOIC; (1) miniature surface mountable electronic packaged component (2) an integrated circuit packaged in a newer, smaller surface mounted container. See Surface Mount Technology, Surface Mounted Device.

Small Outline Transistor SOT; a surface mount package used to house transistors for assembly. The SOT-23 is the most common.

SMD (1) Surface Mounted Device; see full term (2) Storage Module Drive.

SMIF Standard Machine Interface Format; see full term.

SMOBC Solder Mask On Bare Copper.

SMT Surface Mount Technology; see full term.

Snap Off Distance the screen printer distance setting between the bottom of the screen and top of the substrate. See Breakaway.

Snowflakes an adhesive layer defect appearing as a blister; due to trapped solvent or other source of outgas during heating.

Soak Time the length of time a ceramic material (such as a substrate or thick-film composition) is held at the peak temperature of the firing cycle.

Soft Solder a low-melting solder, generally a lead-tin alloy, with a melting point below 800°F (425°C).

SOIC Small-Outline Integrated Circuit; see full term.

Solder metal alloy, usually 62/38 tin/lead, used to coat circuit conductors and/or attach electronic components; can be printed, plated, or roll coated.

Solder Bump the round solder balls bonded to an IC contact area and used to make connection to a conductor by face-down bonding techniques.

Solder Cream solder paste generally used for reflow soldering of electronic components.

Solder Dam a dielectric composition screened across a conductor to limit molten solder from spreading further onto solderable conductors.

Solder Mask a selectively applied ink, coating, or film to prevent solder from coating and adhering to protected areas.

Solder Resist a material used to localize and control the size of soldering areas, usually around component mounting holes. The solderable areas are defined by the solder resist matrix.

Solderability the ability of a conductor to be wet by molten solder and to form a strong bond with the solder.

Soldering the process of joining metals by fusion and solidification of an adherent alloy having a melting point below about 800°F.

Solidus the locus of points in a phase diagram representing the temperature, under equilibrium conditions, at which each composition in the system begins to melt during heating, or complete freezing during cooling.

SOT Small Outline Transistor; see full term.

Spacing in electronic circuits, the distance between conductor edges.

Specific Heat the ratio of the thermal capacity of a material to that of water (2) quantity of heat required to raise the temperature of 1 g of material by 1°C.

Split Tip Electrode same as parallel-gap electrode.

Sputtering the removal of atoms from a source and redeposition onto a target, by energetic plasma ion bombardment. The process is used to deposit very thin films of both metals and nonmetals.

Squeegee neoprene or other elastomeric blade, drawn across a printing screen which forces ink through a selective stencil. See Screen Printing.

Squeeze Out see Adhesive Squeeze Out.

SSWS Static-Safe Work Station; see full term.

Stainless Steel Screen a stainless steel mesh screen stretched across a frame and used to support a circuit pattern defined by an emulsion bonded to the screen.

Stair Step Print a print which retains the pattern of the screen mesh at the line edges. This is a result of inadequate dynamic printing pressure exerted on the paste or insufficient emulsion thickness coating the screen.

Stamped Printed Wire circuitry produced by die stamping out a conductor pattern from a metal foil and bonding to an insulator base.

Standoff a connecting post of metal bonded to a conductor and raised above the surface of the film circuit.

Static Rinse a captive rinse which does not drain. The bath is periodically replaced.

Static Safe Work Station SSWS; a manufacturing or inspection station with static control equipment that reduces or eliminates static discharge which can damage or destroy electronic devices.

Steel Rule Die a simple, low cost blanking die comprising a thin steel ribbon or rule embedded into a wood or metal base so that the protruding edge forms the desired cutting pattern.

Stencil a thin sheet material with a circuit pattern cut into the material. A metal mask is a stencil.

Step that break in a surface intending to locate a mask. Usually a demarcation between two different colors or textures.

Step and Repeat (1) a method by which successive exposures of a single image are made to produce a Multiple Image Production Master (2) a process wherein the printing pattern is repeated many times in evenly spaced rows onto a single film or substrate. This permits the production of several items during each process cycle.

Step Plating rough plating with the appearance of a terrace or steps of electrodeposited metal.

Step Soldering the technique for sequentially soldering connections using solder alloys with different melting temperatures.

Stitch Bond a bond made with a capillary-type bonding tool when the wire is not formed into a ball prior to bonding. Used to connect electronic chips.

Stress-Corrosion Cracking SCC; a degradation of metal, accompanied by fracture, caused by corrosion and accelerated by stress. The term is often applied to printed circuits with the corrosion source being solder flux.

Stress Free the annealed or stress relieved material where the molecules are no longer in tension.

Stress Relieve (1) a process of reheating a film resistor to make it stress free (2) also the incorporation of antistress elements in circuits, especially thin flexible types.

Strip (1) the removal of etch or plating resist after processing (2) the removal of electroplate.

Stripline (1) an electrical impedance controlling configuration where a ground plane is placed either above or below the signal line (2) a microwave conductor on a substrate (3) a type of transmission line consisting of a single narrow conductor parallel and equidistant to two parallel ground planes for the purpose of impedance control.

Substrate (1) the supporting material upon which the elements of a hybrid microcircuit are deposited or attached or within which the elements of an integrated circuit are fabricated (2) the initial base material which is processed to produce a circuit, switch or overlay.

Subtractive Process in circuit manufacturing, a method of producing conductor patterns by selectively removing a previously applied conductive layer such as by chemical etching.

Superconductivity the property of a solid to conduct electricity, the flow of electrons, with virtually no electrical resistance. Superconductivity occurs only at very low temperatures.

Surface Conductance conductance of electrons along the outer surface of a conductor.

Surface Diffusion the high-temperature injection of atoms into the surface layer of a semiconductor material to form the junctions. Usually, a gaseous diffusion process.

Surface Etch a defect appearing as a partially etched, rough etched metal; caused by resist breakdown part-way thru the etching process.

Surface Mounted Assembly SMA; (1) an electronic module or assembly consisting of Surface Mounted Devices bonded to a circuit board (2) Circuits containing surface mounted devices.

Surface Mounted Device SMD; (1) an electronic component housed in a small package that is connected to a circuit by attaching it to the surface with solder or conductive adhesive (2) Miniature electronic devices, without leads, that are attached to the surface of a circuit.

Surface Mount Technology SMT; circuit assembly technology where electronic devices are housed in smaller packages that are mounted on the surface of circuit boards.

Surface Resistivity the resistance of a material between two opposite sides of a unit square of its surface, commonly expressed in ohms per square.

Surface Tension an effect of the forces of attraction existing between the molecules of a liquid. It exists only on the boundary surface.

TAB Tape Automated Bonding; see full term.

TAB-OB Tape Automated Bonding On Board; see full term.

Tacking Plate a fixture used to hold material in position during the application of coverlay film.

Tactile Feel for touch switches, the operator perception through sense of touch, that switch actuation has taken place. A change in depression force occurs with displacement.

Tape Automated Bonding TAB; a fine-line flexible circuit resembling 35mm photo film which is used to bond silicon chips directly onto the circuit by ultrasonic wedge bonding. The TAB is subsequently bonded to a circuit board.

Tape Automated Bonding on Board TAB-OB; same concept as Tape Automated Bonding (TAB), except the chip attachment pads are incorporated into a full circuit. A TAB-OB may incorporate several chips per circuit.

Tape Test an adhesion test where the finish is scribed with a cross-hatch and pressure sensitive tape is applied and pulled away quickly.

Tarnish chemical accretions on the surface of metals, such as sulfides and oxides. Solder fluxes have to remove tarnish in order to allow wetting.

TC Thermocompression (Bonding); see full term.

TCC Temperature Coefficient of Capacitance; see full term.

TCE Temperature Coefficient of Expansion; see Temperature Coefficient of Thermal Expansion.

TCR Temperature Coefficient of Resistance; see full term.

TDR (1) Time Domain Reflectometer; see full term. (2) Time Delay Relay; now computerized with a single chip micro, the TDR permits programming of delays depending on inputs.

Tear Strength measurement of the amount of force needed to tear a solid material that has been nicked on one edge and then subjected to a pulling stress. Measured in lb/inch.

Temperature Aging aging or stressing a film circuit in an elevated temperature over a period of time. The term applies also to any product testing where heat accelerates the aging process.

Temperature Coefficient of Capacitance TCC; a term describing the change in electrical capacitance with change in temperature.

Temperature Coefficient of Expansion see Temperature Coefficient of Thermal Expansion.

Temperature Coefficient of Thermal Expansion TCE; the dimensional change in a material with temperature.

Temperature Coefficient of Resistance TCR; a value describing the change in electrical resistance with change in temperature in a reversible manner.

Temperature Cycling an environmental test where the film circuit is subjected to several temperature changes from a low temperature to a high temperature over a period of time.

Tensile Strength the pulling stress which needs to be applied to a material to break it, usually measured in psi.

Tenting a printed board fabrication method of covering over plated-through holes and the surrounding conductive pattern with a resist, usually dry film.

Terminal a keyboard and display which interfaces with a main computer. Terminals without computing power are called ''dumb.''

Terminal Junction circuit area where all conductors merge; usually applies to automotive instrument cluster circuits.

Test Dry usually after 24 hours, ready for test.

Test Pattern a circuit or group of substrate elements processed on or within a substrate to act as a test site or sites for element evaluation or monitoring of fabrication processes.

T_g Glass Transition Temperature; see full term.

Thermal Conductivity the rate with which a material is capable of transferring a given amount of heat through itself.

Thermal Design the schematic heat flow path for power dissipation from within a film circuit to a heat sink.

Thermal Drift the drift of circuit elements from nominal value due to changes in temperature.

Thermal Drop the difference in temperature across a boundary or across a material.

Thermal Expansion the increase in length of a dimension under consideration caused by an increase in temperature expressed as in./in./deg or PPM/deg.

Thermal Gradient the plot of temperature variances across the surface or the bulk thickness of a material being heated.

Thermal Management application of techniques and design considerations to control and dissipate the heat generated by electronic devices.

Thermal Mismatch differences of thermal coefficients of expansion (TCE) of materials which are bonded together which can result in warpage or fracturing of joints.

Thermal Movement the growth and shrinkage of thermoplastic parts due to changes in ambient conditions; usually in terms of thermal coefficient of expansion.

Thermal Runaway a condition wherein the heat generated by a device causes an increase in heat generated. This spiraling rise in dissipation usually continues until a temperature is reached that results in destruction of the device.

Thermal Shift the permanent shift in the nominal value of a circuit element due to heating effect.

Thermal Shock a condition whereby devices are subjected alternately to extreme heat and extreme cold. Used to screen out processing defects.

Thermocompression a non-adhesive bonding process where two metals, under heat and pressure, form a strong bond.

Thermocompression Bonding a process involving the use of pressure and temperature to join two materials by interdiffusion across the boundary.

Thermoplastic a resin or polymer that can be repeatedly melted or softened by heating without a permanent change in properties.

Thermoset an adhesive, coating, ink or resin which undergoes a chemical cross-linking (solidifies or the mp increases substantially; typically becomes non-melting) by application of heat, UV or other energy, or introduction of catalyst or co-reactant. Examples: phenolics, urethanes, epoxies. In a precise definition, non-thermal energy is excluded, but the term is now applied to all highly cross-linked plastics. See Thermoplastic.

Thermosonic Bonding bonding combining thermocompression and ultrasonic bonding. Typically, gold wire is bonded to integrated circuit.

Thermoswaging heating a pin that is inserted in a hole and upsetting the hot metal so that it swells and fills the hole, thereby forming a tight bond with the base material.

Thick Film a film deposited by screen printing processes and fired at high temperature to fuse into its final form. The basic processes of thick-film technology are screen printing and firing.

Thick Film Circuit a microcircuit in which passive components of a ceramic-metal composition are formed on a suitable substrate by screening and firing.

Thick Film Direct Process same as Thick Film Direct Writing.

Thick Film Direct Writing a computer-controlled (CAD/CAM) thick film orifice-printing or ink-jet printing technology for direct writing of film circuitry at high speeds. Conductors, resistors and other elements can be applied.

Thief a rack for electroplating that provides more even current density; the design absorbs uneven current distribution on irregular parts.

Thin Film a thin film (usually less than 100 microns) is one that is deposited onto a substrate by an accretion process such as vacuum evaporation, plasma sputtering or pyrolytic decomposition.

Thixotropic a rheological property where a relatively thick material becomes thinner on agitation. Also called Shear Thinning.

Three Layer Tab an electronic interconnecting medium for microchips consisting of copper circuitry/adhesive/polyimide film. An access window is pre-punched in the film. Also see Tape Automated Bonding (TAB).

Time Domain Reflectometer TDR; an analytical instrument.

Tinned literally, coated with tin, but commonly used to indicate coated with solder.

Tinning to coat metallic surfaces with a thin layer of solder.

TO Package abbreviation for transistor outline, established as an industry standard by JEDEC of the EIA.

Tombstoning a soldering defect phenomenon in Surface Mount Assembly of rectangular chip packages where one end of a component does not bond, and consequently, the high surface tension at the good solder joint tips the component upright giving the appearance of a miniature tombstone.

Tooling Plate the vacuum plate on a flat bed screen printer; consists of drilled holes and a manifold below to connect vacuum to the holes; holds down substrate.

Topography the surface condition of a film; bumps, craters, etc.

Topology the surface layout, design study and characterization of a microcircuit. It has application chiefly in the preparation of the artwork for the layout masks used in fabrication.

Torque Test a test for determining the amount of torque required to twist off a lead or terminal.

Touch Screen a computer input device consisting of a touch-sensitive screen that is location responsive. A menu appearing on a screen can be activated by touching the corresponding screen area.

Trace a conductor appearing as metal ribbon or printed wire.

Track same as trace but more commonly used in Europe.

Tracking the ability of two similar elements on the same circuit to change values with temperature in close harmony.

Transfer Bumped Tab tape automated bonding material with bumps for connecting to the integrated circuit, produced by a transfer method. Bumps are formed first on glass and then transferred to the TAB.

Transverse Direction also called TD—across the web. Right angles to machine direction.

Travel the amount of displacement which must occur in a membrane or button type switch for actuation.

Treeing a tree-like electroplating normally found at the edge of a circuit. Caused by too high a current density.

Triboelectric pertaining to or the production of electricity by friction.

Triboelectric Charging generation of static electricity by two substances rubbing together.

Trim Notch the notch made in a resistor by trimming to obtain the design value. See Kerf.

Trimming notching a resistor by abrasive or laser means to raise the nominal resistance value.

TS Thermosonic (Bonding); see full term.

UL Underwriters Laboratories, Inc., see full term.

Ultrasonic (1) vibrational energy above the audible range ($>20{,}000$) and used to weld plastics and metal, clean materials or perform other mechanical operations. (2) a bonding method for plastics wherein vibrational energy heats the bonding surface to a melt. (3) a bonding method for metal, especially wire, used to connect electronic chips to circuits and lead frames.

Ultrasonic Bonding a process involving the use of ultrasonic energy and pressure to join two materials.

Ultrasonic Cleaning a method of cleaning that uses cavitation in fluids caused by applying ultrasonic vibrations to the fluid.

Ultraviolet UV; an energetic form of "light," just beyond visible purple, having sufficient energy to produce ionization in certain materials, capable of polymerizing UV curable monomers, and of inducing chemical reactions. UV can also degrade organic inks, colorants and plastic films.

Undercut a condition where etchant removes too much metal from the sides of the metal being etched resulting in overhang at the top of the metal where the resist is in place.

Underwriters Laboratories, Inc. UL; a private corporation which tests and approves manufactured goods, especially electrical.

Underwriters Symbol a logotype authorized for placement on a product which has been recognized (accepted) by Underwriters Laboratories, Inc. (UL).

Unit Under Test UUT; term applied to any circuit board which is being tested by ATE.

US Ultrasonic; see full term.

UUT Unit Under Test; see full term.

UV Ultraviolet; see full term.

UV Covercoat see Covercoat and UV Dielectric.

UV Dielectric a UV (ultraviolet radiation) curable electrically insulating screening ink applied over conductors for protection or to build cross-overs.

Vacuum Autoclave Lamination VAL; a circuitry manufacturing process where a multilayer circuit stack, sealed within a plastic vacuum "turkey bag," is laminated together in an autoclave which supplies the heat and pressure. This process is considered an improvement over VHL.

Vacuum Deposition deposition of a metal film onto a substrate in a vacuum by metal evaporation techniques.

Vacuum Hydraulic Lamination VHL; a circuitry manufacturing process where a multilayer circuit stack is laminated in a hydraulic platen press which is within a vacuum chamber. Alternatively, vacuum is applied to plastic "turkey bags" within a hydraulic press.

Vacuum Pickup a handling instrument with a small vacuum cup on one end used to pick up chip devices.

VAL Vacuum Autoclave Laminating; see full term.

Vapor Degreaser a common apparatus for cleaning microelectronic parts and assemblies. Basically a tank of boiling solvent with a vapor zone maintained above it combined with means for immersion, spray, and rinse.

Vapor Phase Reflow the reflowing of a material, especially solder, by immersion in hot vapors. Fluorochemical liquids are generally used because of their high heat capacity which permits rapid heating of material to be reflowed.

Vapor Phase Soldering VPS; a reflow solder method using condensing hot vapors as the heat transfer means.

Varnish a protective coating for a circuit to protect the elements from environmental damage.

Vehicle a thick-film term that refers to the organic system in the paste.

Via(s) (1) openings in the insulator layer(s) of multi-layer thick film circuits through which conductor is printed for inter-layer connection. (2) an opening in the dielectric layer through which a riser passes. (3) vertical conductor or conductive path following the interconnection between multilayer circuit layers.

Viscosity a term used to describe the fluidity of material, or the rate of flow versus pressure. The unit of viscosity measurement is poise, more commonly centipoise. Viscosity varies inversely with temperature.

Viscosity Coefficient the value of the tangential force per unit area which is necessary to maintain a unit relative velocity between two parallel planes a unit distance apart.

Viscometer a device that measures viscosity. Viscometers for thick-film compositions must be capable of measuring viscosity under conditions of varying shear rates.

Volt (E) or (V); unit of electromotive force; potential difference required to make a current of 1 ampere flow through 1 ohm of resistance.

Voltage Gradient the voltage drop (or change) per unit length along a resistor or other conductance path.

Voltage Rating the maximum voltage which an electronic circuit can sustain to ensure long life and reliable operation.

Volume Resistivity (1) a 1 cm cube of material will have resistance equal to the material's resistivity. The qualification "volume" adds nothing, but is sometimes used so that "resistivity" and "resistance" will not be confused. (2) electrical resistance between opposite faces of a 1 cm cube of material commonly expressed in ohm centimeters (ohm cm).

VPS Vapor Phase Soldering; see full term.

W Band a microwave radio band from 56–100 gigahertz.

Waffle Pack an open, compartmentized container for holding dies or surface mount devices for loading by automatic equipment. See also Dry Pack.

Warp and Woof threads in a woven screen which cross each other at right angles.

Warpage the distortion of a substrate from a flat plane.

Wave Soldering a component attachment process where a circuit assembly is brought into contact with a molten wave of solder so that component connections are soldered to the circuit board.

Waviness one or a series of elevations or depressions or both, which are readily noticeable and which include defects such as buckles, ridges, etc.

WDX Wavelength Dispersion X-Ray; see full term.

Wedge Bonding a bond made with a wedge tool. The term is usually used to differentiate thermocompression wedge bonds from other thermocompression bonds. (Almost all ultrasonic bonds are wedge bonds.)

Wetting the spreading of molten solder on a metallic surface, with proper application of heat and flux.

Wicking (1) migration of liquid, by capillary action, along base material (2) plating underneath poorly adhering resist (3) the flow of solder along the strands and under the insulation of stranded lead wires.

Wire Bond includes all the constituent compounds of a wire electrical connection such as between the terminal and the semiconductor die. These components are the wire, metal bonding surfaces, the adjacent underlying insulating layer (if present), and substrate.

Wire Bonding a method of connecting silicon or other electronic chips to carriers or to circuits by means of very fine wires which are thermosonically or ultrasonically bonded, usually by automatic equipment.

Wiring Crossover Jumper glass jumpers with two or more gold-metallized bonding areas (pads). They are eminently suitable for use as wiring crossovers in prototype hybrid fabrication where special conductor patterns may be realized without the need for a layout of and work on extra thick film screening/firing steps.

Wobble Bond a thermocompression, multicontact bond accomplished by rocking (or wobbling) a bonding tool on the beams of a beam lead device.

Woven Screen a screen mesh used for screen printing usually of nylon or stainless steel or possibly silk.

YAG Yttrium Aluminum Garnet; see full term.

Yellow Room colloquial name in the industry for a room or an area illuminated solely by yellow lighting, usually yellow fluorescent bulbs. The purpose is to eliminate any ultraviolet rays.

Yield the ratio of usable components at the end of a manufacturing process to the number of components initially submitted for processing. Can be applied to any input-output stage in processing, and so must be carefully defined and understood.

Z-Axis Conductive Adhesive an adhesive dry film or liquid which conducts electrically in only the vertical, or Z-axis, permitting electrical connections to be made between two circuits or between a circuit and an electronic element.

Zebra Strip a material consisting of stacked layers of conductor and insulator, so when placed on edge, interconnection can be made between a circuit and a device such as a glass electronic display. The name derives from the appearance due to dark conductor (carbon) and light insulator.

Zippers diagonal wrinkles with puckering so as to look like a zipper.

Index